LOW-ENTHALPY GEOTHERMAL RESOURCES FOR POWER GENERATION

Low-Enthalpy Geothermal Resources for Power Generation

D. Chandrasekharam
Department of Earth Sciences,
Indian Institute of Technology Bombay, India

Jochen Bundschuh
International Technical Cooperation Program, CIM (GTZ/BA), Frankfurt,
Germany—Instituto Costarricense de Electricidad (ICE), San José, Costa Rica
Royal Institute of Technology (KTH), Stockholm, Sweden

CRC Press
Taylor & Francis Group
Boca Raton London New York Leiden

CRC Press is an imprint of the
Taylor & Francis Group, an **informa** business

A BALKEMA BOOK

CRC Press/Balkema is an imprint of the Taylor & Francis Group, an informa business

© 2008 Taylor & Francis Group, London, UK

Typeset by Vikatan Publishing Solutions (P) Ltd., Chennai, India
Printed and bound in Hungary by Uniprint International bv (a member of the Giethoorn Media-group), Székesfehérvár

Published by: CRC Press/Balkema
      P.O. Box 447, 2300 AK Leiden, The Netherlands
      e-mail: Pub.NL@taylorandfrancis.com
      www.crcpress.com – www.taylorandfrancis.co.uk – www.balkema.nl

Library of Congress Cataloging-in-Publication Data
Low-enthalpy geothermal resources for power generation / by D. Chandrasekharam and Jochen Bundschuh.
     p. cm.
  Includes bibliographical references and index.
  ISBN 978-0-415-40168-5 (hardcover) -- ISBN 978-0-203-89455-2 (ebook)
1. Geothermal engineering 2. Geothermal resources. I. Chandrasekharam, D. II. Bundschuh, Jochen.

TK1055.L68 2008
621.44 – dc22

                            2008003160

ISBN: 978-0-415-40168-5 (hbk)
ISBN: 978-0-203-89455-2 (ebook)

Table of Contents

Foreword

Today majority of the countries depend on imports of fossil fuels to secure their energy and developmental activities. The global proven oil reserves were estimated to be around 137 trillion liters at the end of 2004 and expected to last for another 40 to 50 years. This security will not last for long and developing countries have increasingly to compete with the developed world in future for fossil fuel. The best option available for the developed countries, and developing countries as well, is to have an energy source mix and reduce dependency on fossil fuels. Non-conventional energy sources play an important role in making these countries more energy independent. Developing non conventional energy has two fold advantages: (1) makes the country energy independent and (2) reduces the CO_2 emission and secures the environment for the future generations. Low-enthalpy geothermal energy is one such source that the developing countries are looking as an option to be energy independent.

From a modest start of 10 kW in 1904 in Larderello Italy, geothermal energy to day is generating greater than 8900 MWe in 25 countries supplying 56,831 GWh/year, operating with an average capacity factor of 73% and producing power online 97% of time. About 1% of the world's population is currently using this energy today. The current production is from established high-enthalpy systems but a large volume of low-enthalpy resources, whose potential is much larger, is lying unutilized in many developing and developed countries. This energy needs to be exploited to make the countries move towards energy independence.

The greenhouse gas emissions are unevenly distributed between the regions of the world. The total greenhouse gas emitted by all the countries is about 6234 million tonnes carbon dioxide per year. Asia and Pacific region contributes maximum amounting to 2167 million tonnes carbon dioxide per year. India and China in the Asian region are the maximum emitters while the North American countries contribute substantially from the Pacific region. Some of the developing countries have already entered in to carbon trade with developed countries to secure their energy supply. This may help in reducing the carbon dioxide emission but does not give energy independence to these countries.

The current book is being published at a time when all the countries are carving for more electricity and be energy independent. The authors unveiled all the provinces where low-enthalpy geothermal resources are lying untapped and contribute substantially to the future electricity demand. Both exploration techniques and economics of power production from small power plants to meet rural electricity demand are well documented in the book. Cost comparison between diesel generated power and geothermal power demonstrates the benefits to the rural areas in terms of cost, socio-economic values, rural employment and preservation of the environment by using this non conventional energy source.

I complement the authors for their contribution to improve the energy security of developed and developing countries and to global climate change mitigation. I am sure that this book will be useful not only for the students but also to researchers, energy planners and policy makers.

Prof Ashok Misra
Director, IIT Bombay, India

Authors' preface

The geothermal resources of the earth are huge, with low-enthalpy resources (<150 °C) having a manifold larger potential, and much wider regional distribution compared to high-enthalpy resources. High-enthalpy resources are presently used for power generation in 16 countries with a total installed capacity of 9000 MW_e. The available resources are vast compared to the total net electricity demand, which is expected to double in the next three decades, with the highest world demand rate expected in the developing countries. Only a small part of this geothermal energy is being extracted economically using existing technology. Geothermal resources are also much larger compared to all fossil fuel resources put together. Compared to other renewables, geothermal resources allow for a much more efficient and stable power supply. This energy is freely available to humanity, and the method to harness it for the betterment of humanity lies in the hands of those who need it. The technology for low-enthalpy geothermal resources is growing at so fast a rate, that in the near future non-renewable fossil fuels may be obsolete! Continuous development of innovative drilling and power generation technologies especially, makes low-enthalpy geothermal resources the best option available to meet the required future electricity demand. It will also guarantee energy security and energy independence for both developing and developed countries, while at the same time drastically reducing greenhouse gas emissions, and thus mitigating global climate change. Antagonists may argue that geothermal sources alone may not meet all the future energy demands of the developed and developing world. This may be true in the short run, but in such situations mixing energy sources is a viable option to meet future electricity demands and mitigate climate change. In the long run, when enhanced geothermal system techniques become more commercial, all the countries may leave CO_2-free atmospheres behind for posterity to enjoy. If the MIT 2006 report has any value, this will be fact and not fiction.

A report published in the Finance and Development report of the World Bank in 1997 states, *"Energy markets do not function efficiently in many developing countries, particularly in rural areas, where nearly 2 billion people do not have electricity . . . Inadequate energy markets threaten to dampen economic growth, hobble development, and keep living standards low"*. Seven years after this statement was made, the scenario had not changed in the rural sector. This is evident from the Annual Meeting address by the World Bank President in 2004, that reads, *"We must give higher priority to renewable energy. New and clean technologies can allow the poor to achieve the benefits of development without having to face the same environmental costs the developed world has experienced"* The current situation (2008) is no better than what rural developing countries were in a decade ago. Today, 90% of the world population living in rural areas in developing countries has no way to meet basic needs like nutrition, heating, and light in spite of the fact that technologies for developing renewable energies have jumped by leaps and bounds, especially with respect to geothermal energy.

Generating electricity from low- and high-enthalpy geothermal energy resources could provide uninterrupted power supply to these rural masses. Case studies described in this book clearly demonstrate how low-enthalpy geothermal resources can be utilized for improving the socio-economic status of rural areas in developing countries. Additional industrial applications such as greenhouse cultivation, space heating, using the spent fluids from heat exchangers, etc., will further uplift the economic status of the rural population by creating employment.

A shortage of trained manpower for exploration and exploitation programs related to geothermal energy resources could be a future problem, especially in the developing countries. A cursory glance at the recent country updates in the Proceeding of the World Geothermal Congress 2005, reveals that allocation of manpower for developing geothermal resources is not only far below that required but also insufficient in most developed countries. There is a lack of and urgent need for easily understandable and accessible information in a comprehensive form on low-enthalpy geothermal resources practically available all over the world. No doubt, such information does exist in several published scientific papers, but, this information is difficult for graduate students, researchers and decision makers, to collect within a short period of time. This is especially true in the case of students in developing countries, who face difficulties accessing such information. This, together with the absence of a comprehensive book that deals with all aspects of low-enthalpy resources, including occurrence, exploration methods, technologies, economics, and global climate change mitigation potential, inspired us to write this book.

This book explains the occurrences of low-enthalpy systems lying unutilized in both developed and developing countries, and their vast potential in different regions and large economies for power generation.

After a general introduction we address, in Chapter 2, the electricity demand and source mix forecasts for future decades in different regions and large economies, and the role of geothermal resources in the 2004–2030 scenario for power generation.

Chapters 3 and 4 deal with the distribution and potential of low-enthalpy resources guaranteeing energy security and independence for both developing and developed countries, while at the same time reducing CO_2 and other greenhouse gas emissions.

In Chapter 5, geographic distribution of low-enthalpy resources is given together with their geological, tectonic, geochemical, and geophysical characteristics. To understanding both wet and enhanced geothermal systems, a sound concept of the geological and tectonic features that control the geothermal systems needs to be understood. With this in mind, we have tried to bring major geothermal provinces associated with different tectonic regimes around the world together and explain all the possible sources where low-enthalpy geothermal energy resources occur for future development.

In Chapters 6 and 7, geochemical and geophysical exploration methods are discussed. Simple geochemical and geophysical methods are essential in understanding the systems during pre-drilling stages. The most expensive component in geothermal power development is the drilling. The cost can be reduced by applying the above methods to understand the geothermal reservoir conditions and to better locate sites for exploratory drill holes.

Chapters 8 and 9 deal with available power generation techniques, and the economic aspects of low-enthalpy geothermal resources. Power generation techniques are important in geothermal energy development programs. In the present competitive world providing affordable electricity to the rural masses without government subsidy, is a challenge. We have shown how this is possible by using a free energy source such as geothermal, which is independent of fluctuations in oil, coal, and other energy source materials' cost. The initial cost of developing geothermal power projects may be higher than those for fossil fuel based plants, but the advantages of using geothermal sources are quite large because these costs are absorbed in the system itself. For example, unit cost of electric power is low in the case of coal-based power plants, but it has hidden cost in the form of environmental mitigation. In the end, the consumer has to pay for both the electric power as well as cleaning the environment. But in the case of geothermal, there is no hidden cost, the investment is made one time, and the unit cost of power, unlike fossil fuel based power, does not fluctuate with time and space.

Finally, in Chapter 10 we discuss the potential of low-enthalpy geothermal resources for rural electrification and give some case studies for small-scale power plants. Electric sector reforms are transforming the potential owners and operators of small geothermal projects from public utilities to private power producers. Reforms are intended to improve the overall economic efficiency of the electric sector and may open new opportunities for small geothermal projects in this more competitive market. The economic analysis together with the case studies described in Chapters 9 and 10 should interest the government officials involved in drawing electricity reforms in developing countries. Future technological developments in terms of drilling, heat exchangers, and binary fluids add additional advantage to low-enthalpy geothermal resources development. Such technologies will place the geothermal resources at the top of the energy ladder. Developing countries have immense opportunity to increase their country's GDP and develop economic growth by becoming energy-independent countries in the next decade through geothermal energy resources.

The comprehensive integral approach of low-enthalpy geothermal resources in this book makes it a convenient source that aims at developing a strong human resource base in developing and developed countries. Developing countries have a good opportunity to overcome their current and future electricity demands by utilizing this resource. The book is intended not only for graduate and research students as a primary dictionary, but also should prove useful for professional geologists, engineers, and people involved in energy planning and greenhouse gas mitigations. The book also addresses members of pertinent national, regional, and international communities involved in energy and climate change mitigation issues. It is a useful information source for decision and policy making, for administrative leaders both in governments and in international bodies such as the United Nations family, the international and regional development banks, financial institutions, and donors, concerned with technical and economic cooperation with developing countries. As authors, we hope that this book will be useful for many people helping society to effectively use the huge available low-enthalpy geothermal resources, providing energy security and energy independence to their countries, and thus contribute to global climate change mitigation.

D. Chandrasekharam
Jochen Bundschuh

About the authors

Dornadula Chandrasekharam (1948, India) has been a Professor in the Department of Earth Sciences, Indian Institute of Technology Bombay (IITB), since 1987. Currently he is the Head, Centre of Studies in Resources Engineering. He obtained his MSc in Applied Geology (1972) and PhD (1980) from IITB. He has been working in the fields of volcanology, groundwater pollution, and geothermics for the past 25 years. Before joining IITB he worked as a Senior Scientist at the Centre for Water Resources Development and Management, and Centre for Earth Science Studies, Kerala, India for 7 years. He was a Third World Academy of Sciences (TWAS, Trieste, Italy) Visiting Professor to Sanaa University, Yemen Republic between 1996–2001, and a Senior Associate of Abdus Salam International Centre for Theoretical Physics, Trieste, Italy from 2002–2007. He received the International Centre for Theoretical Physics (ICTP, Trieste, Italy) Fellowship to conduct research at the Italian National Science Academy (CNR) in 1997. Prof. Chandrasekharam extensively conducted research in low-enthalpy geothermal resources in India and is currently the Chairman of M/s GeoSyndicate Power Private Ltd., the only geothermal company in India. He is a member of the International Geothermal Association, and has widely represented the country in several international geothermal conferences. He conducted short-term courses on low-enthalpy geothermal resourses in Argentina and Costa Rica. He has supervised 17 PhD students and published 73 papers in international and 35 papers in national journals of repute. He is the Editor of "Geothermal Energy Resources for Developing Countries" (2002) and "Natural Arsenic in Groundwater" (2005) published by AA Balkema Publishers. He is one of the executive members of the International Society of Groundwater for Sustainable Development (ISGSD).

Jochen Bundschuh (1960, Germany), finished his PhD on numerical modeling of heat transport in aquifers in Tübingen in 1990. He is working in geothermics, subsurface and surface hydrology and integrated water resources management, and connected disciplines. From 1993 to 1999 he served as an expert for the German Agency of Technical Cooperation (GTZ) and as a long-term professor for the DAAD (German Academic Exchange Service) in Argentine. In 2001 he was appointed to the Integrated Expert Program of CIM (GTZ/BA), Frankfurt, Germany and works within the framework of the German governmental cooperation as adviser in mission to Costa Rica at the *Instituto Costarricense de Electricidad* (ICE). Here, he assists the country in evaluation and development of its huge low-enthalpy geothermal resources for power generation. In 2005, he was appointed as an affiliate professor of the Royal Institute of Technology, Stockholm, Sweden.

Prof. Bundschuh is the editor of the books "Geothermal Energy Resources for Developing Countries" (2002), "Natural Arsenic in Groundwater" (2005), the two-volume monograph "Central America: Geology, Resources and Hazards" (2007), and the author of the book "Modeling of Groundwater and Geothermal Systems" (2008), all by A.A. Balkema Publishers and the Taylor and Francis Group. He is also the author of over 70 international scientific publications. In 2006, he was elected Vice-President of the International Society of Groundwater for Sustainable Development.

Acknowledgements

This book would be incomplete without an expression of our sincere and deep sense of gratitude to Prof. S. Viswanathan (India) for editing the entire manuscript and making necessary corrections. We appreciate his patience and willingness in improving the contents of the book. We thank geologist Gerardo Soto (Costa Rica) for his critical review and constructive comments. Many thanks also to Sarah Pasela (Canada), for devoting valuable time improving the English language. We wish to express our sincere thanks to them, whose efforts contributed to the high quality of the book.

Jochen Bundschuh is especially grateful to the support provided by the Integrated Expert Program of CIM*, Frankfurt, Germany, and for delegating him as long-term integrated expert to the *Instituto Costarricense de Electricidad* (ICE) in San José, Costa Rica. This mission allowed, within the framework of the German governmental cooperation in the area of geothermal resources, the compilation of this book, which will remain as a strong outcome of this cooperative effort between the governments of Germany and Costa Rica, contributing to a worldwide extensive use of high- and low-enthalpy geothermal resources. Many thanks to the *Instituto Costarricense de Electricidad* (ICE), for supporting the compilation of this book.

Additionally, we would like to thank NASA for their courtesy in permitting the use of their digital elevation models PIA3388 (World), PIA3377 (North America), and PIA3364 (Central America) (http://photojournal.jpl.nasa.gov) [Courtesy NASA/JPL-Caltech]. These regional digital elevation models of 90 m spatial resolution, which were obtained from the Shuttle Radar Topography Mission (STRM) of NASA, are excellent for assessing larger-scale morphology and were, therefore, used in many of the figures in the book as backgrounds for the maps.

D. Chandrasekharam
Jochen Bundschuh

---

* Auftraggeber: Bundesministerium für Wirtschaftliche Zusammenarbeit und Entwicklung - (BMZ), Centrum für internationale Migration und Entwicklung (CIM). [Federal Ministry for Economic Cooperation and Development (BMZ), Center for International Migration and Development (CIM)].

CHAPTER 1

Introduction

"The latest studies show that it is cheaper to invest in climate protection than to pay for the losses that result from inactivity. It is thus prudent to act now from an economic perspective as well."

T. Jeworrek: United Nations Environment Programme's Finance Initiative, speaking on the occasion of World Environment Day (WED), Oslo, 2007.

Security for long-term electricity supply and greenhouse gas emission from fossil fuel based power plants is a cause of concern for the entire world today. The world net electricity demand is going to increase by a factor of two from 2004 to 2030, rising from the present 16,424 TWh to 30,364 TWh in 2030 (EIA 2007). This demand is expected to grow at the rate of 1.3% per year in OECD (Organization for Economic Cooperation and Development) countries while in non-OECD countries it is expected to grow at the rate of 3.5% per year. This indicates that the demand will be greater in developing countries where electricity generation will increase by several fold relative to industrialized countries. Thus, developing countries are the future primary emitters of CO_2 and other greenhouse gases from fossil fuel based power plants. The reason for such demand in developing countries is due to higher economic and population growth rates, while developed countries may experience only moderate economic growth (Fig. 1.1a). It is very important for developing countries to be judicious in planning for future energy needs and cautiously avoid economic, social, and environmental problems. However, it is clear that the per-capita primary energy and electricity consumption, which are positively correlated to the per-capita GDP, and therefor the per-capita CO_2 emissions, are much lower in the developing world compared to industrialized countries (Fig. 1.1b).

According to the 2004–2030 forecasts for energy mixes, coal and natural gas will continue to be the most important fuels for generating electric power. This confirms that the power sector will remain the single main contributor to global CO_2 emissions, followed by the transport sector. The present and future challenges are to control CO_2 emissions and enhance clean development mechanism (CDM) in order first, to generate electric power through environmentally friendly energy sources, thereby reducing the dependency on coal, and second, to implement technological and infrastructural mitigation options to reduce greenhouse gas emissions from the transport sector. Unless these two goals are achieved, it will be difficult to control global climate change and save future generations from its harmful effects.

Countries that do not have indigenous fossil fuel resources to support their demand for electricity will increase fossil fuel imports. Such countries' fuel supply security and cost of electric power depend on fossil fuel import costs and other uncertainties. On the other hand, countries with enough fossil fuel resources will more aggressively deplete their resources due to their accelerated use of fossil fuel to meet their growing electric power demand.

The above two challenges, i.e. guaranteed electricity supply to meet ever increasing demand, and reduction of CO_2 and other greenhouse gas emissions are the primary goals to be addressed by both developed and developing countries. Nuclear energy is a potential solution to meet electricity demand and reduce CO_2 emissions, but increasing contamination concerns in the world are a hindrance to its future growth. Hence, the only options remaining are using renewable sources of energy. Amongst all the available renewables we need to examine which one can sustain long-term future needs. The hydroelectric power sector is already developed in several industrialized

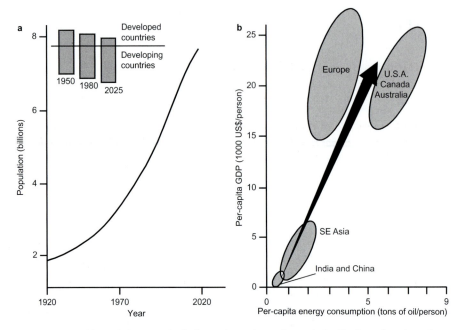

Figure 1.1.   (a) World population growth; the insert shows the north–south distribution, where it can be recognized that developing countries register a steep population growth compared to population in the industrialized world in the next two decades; (b) Economic growth and primary energy demand of developed and developing countries and regions (year 1990) (modified from Afgan and Carvalho 2002).

countries and does not show possibilities of any further development. In many developing countries however, further development of hydroelectric power resources is possible. Growing environmental concerns such as deforestation, displacement of people, acquiring land, and changes in climatic events (changes in weather and rainfall pattern, droughts, etc.) are major factors challenging the future development of this venture. Solar and wind energy are other renewable options, but these sources can only meet part of the energy demand. In the case of wind power, for a constant supply of electric power the system needs a backup system that could operate on diesel. Biomass is the only option that could contribute to meet a major part of the electricity demand in such a case.

To meet future energy demands renewable energy sources should meet the following criteria: (1) the sources should be large enough to sustain a long-lasting energy supply to generate the required electricity of the country, (2) the sources should be economically and technically accessible, (3) the sources should have a wide geographic distribution, and (4) the sources should be low CO_2 emitters to make significant contribution to global warming mitigation.

Low-enthalpy energy resources ($<150\,°C$) meet all the above criteria. This source is huge, is not utilized to the extent that it can be used to provide for the future electricity demand, and is sustainable for generations. Although this source could be developed in the following decades to meet the entire world electricity demand, this source should be mixed with other renewables to diversify electricity generation sources, as detailed in the following chapters.

However, it will not be enough to consider only the low-enthalpy geothermal resources for electricity generation as it is used today; it also needs to be used to partly replace the use of fossil fuels in sectors such as transportation, where fossil fuels can be replaced by electricity-driven motors in public, commercial, and private vehicles.

Keeping in view the four major criteria mentioned above the low-enthalpy geothermal resources for electricity generation are thoroughly discussed in this book.

CHAPTER 2

World electricity demand and source mix forecasts

"World electricity generation nearly doubles in the IEO2007 reference case from 2004 to 2030. In 2030, generation in the non-OECD countries is projected to exceed generation in the OECD countries by 30 percent."

EIA: International Energy Outlook (IEO) 2007.

2.1 WORLD OVERVIEW

As described in Chapter 1, world electricity production is expected to reach 30,364 TWh by the year 2030 (EIA 2007) (Fig. 2.1a). Major demand and electricity production will increase in developing countries since these countries have high future potential economic and population growth rates accompanied by an increase in the standard of living and demand for consumer goods (Fig. 2.1a, Table 2.1). As a consequence, the anticipated electricity demand will be highest in residential and commercial sectors (Fig. 2.1b). Thus, it is important for developing countries to take special care regarding their energy planning in order to avoid economic, social, and environmental problems. The forecast for the mix of energy sources shows that coal and natural gas will remain the most important fuels for electricity generation throughout the 2004–2030 projection period, totaling 80% of the total increment in world electric power generation (Fig. 2.1c). This emphasizes that the electricity sector will remain the main contributor of global CO_2 emissions, followed by the transport sector. Therefore, our goals for the present and following decades must be (1) to substitute coal in the power generation sector with an environmentally friendly energy source, and (2) to implement technological and infrastructural mitigation options to reduce greenhouse gases from the transport sector. If these challenges are not tackled, then global climate change can not be mitigated.

2.2 REGIONAL ELECTRICITY MARKETS AND FORECASTS UNTIL 2030

A short overview on the energy source mixes, the electricity generated at present, and projections for the year 2030 for different countries and regions based on the International Energy Outlook 2007 (IEO 2007) report (IEO reference case; EIA 2007) is given below; supporting data are given in Figures 2.1d, 2.2 and 2.3 and Table 2.2.

OECD North America: In the north American region (USA, Canada, and Mexico), the electricity generation is expected to grow at an average rate of 1.5% per year, from 4043 TWh in 2004 to 7197 TWh in 2030. Although USA is the largest electricity generator, its generation capacity is expected to grow only at an average rate of 1.4% per year (3975 TWh in 2004 to 5797 TWh in 2030). This slow increase of power generation compared to the fast increases of electricity generation in the non-OECD countries, especially those of Asia, results in an reduction of the USA's share in the world electricity generation (2030: 19%) making it the second largest power generator (after China with 21%) compared to the year 2004 when USA was the highest electricity generator (24%). Canada follows the trend of the USA having a low projected increase rate of 1.5% per year for the 2004–2030 period, whereas the Mexican electricity generation growth rate is much higher since its electricity market is not as mature as that of the USA and Canada, which is in agreement with the high economic and demographic growth in this country, and is similar to many non-OECD countries.

OECD Europe: Due to mature electricity markets and a slow population growth, Europe has the slowest growth rate of electricity generation in the world, at an average projected growth rate of

3

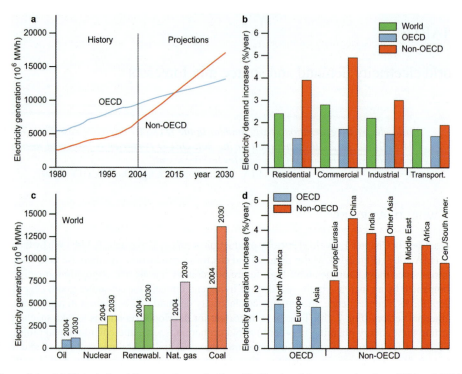

Figure 2.1.   (a) Historical and forecasted growth of worldwide electricity generation from 1980 to 2030 in OECD and non-OECD countries; (b) Electricity demand increases with use by the world, OECD, and non-OECD regions; (c) Electricity sources in 2004 and 2030; (d) Electricity generation increase forecasts for world regions and selected countries (data from EIA 2007).

0.8% per year during 2004–2030. The worldwide share of OECD Europe's electricity generation was 20% in 2004 and is expected to decrease to 13% in 2030.

OECD Asia: The electricity generation in OECD Asia is projected to increase by a rate similar to that of OECD North America at an average of only 1.4% per year during the projected period of 2004–2030 (2004: 1586 TWh; 2030: 2259 TWh) corresponding to a decrease from 10% of the world power generation in 2004 to 7% in 2030. Thus, the countries with mature electricity markets show the lowest increase (Japan: 1.0%; Australia/New Zealand: 1.4%), whereas the electricity generation in the emerging economy of South Korea is expected to grow at an average rate of 2.3% in the 2004–2030 period.

Non-OECD Europe/Eurasia: In non-OECD Europe/Eurasia the electricity generation is fore-casted to grow during 2004–2030 period moderately from 1496 TWh in 2004 to 2731 TWh in 2030, which corresponds to an average increase of 2.3% per year. Russia is expected to remain the largest power generator in this region (59% in 2004, 55% in 2030) (Fig. 2.3). On a worldwide scale, the non-OECD Europe/Eurasia region generates about 9% in both 2004 and 2030.

Non-OECD Asia: In non-OECD Asia the most remarkable relative and absolute changes of electricity generation growth are found. Here the electricity generation is expected to increase from 3518 TWh in the year 2004 to 10,186 TWh in 2030, corresponding to an annual average increase of 4.2%. In this region and worldwide, China and India have by far the highest absolute national electricity generation growths (Figs. 2.1d and 2.3). Thus, in China total electricity generation is expected to increase from 2080 TWh in 2004 to 6338 TWh in 2030, corresponding to a world share of 13% in 2004 to 21% in 2030. This will make China the largest electricity generator in the world (in 2004 this place was occupied by USA). In India the corresponding increase of power generation is forecasted to be from 631 TWh in 2004 to 1704 TWh in 2030, shifting this country

Table 2.1. Population and gross domestic product for the year 2004 and projection for 2030 (data from EIA 2007).

	Population			Gross domestic product (GDP) in year 2000 US$									
				Expressed in market exchange rates					Expressed in purchasing power parity				
				Per person		Total			Per person		Total		
	2004	2030	Increase	2004	2030	2004	2030	Increase	2004	2030	2004	2030	Increase
	10^6	10^6	%/year	US$	US$	10^9 US$	10^9 US$	%/year	US$	US$	10^9 US$	10^9 US$	%/year
OECD	1163	1300	0.4	23650	38814	27505	50458	2.4	24608	41896	28619	54465	2.5
North America	432	537	0.8	28056	47493	12120	25504	2.9	29456	50063	12725	26884	2.9
USA	294	365	0.8	36408	61627	10704	22494	2.9	36408	61627	10704	22494	2.9
Canada	32	39	0.8	24969	37256	799	1453	2.3	31406	46897	1005	1829	2.3
Mexico	106	133	0.9	5830	11707	618	1557	3.6	9585	19248	1016	2560	3.6
Europe	532	562	0.2	17673	28317	9402	15914	2.0	20925	35432	11132	19913	2.3
Asia	199	202	0.0	30060	44752	5982	9040	1.6	23925	37965	4761	7669	1.9
Japan	128	123	−0.2	38023	52626	4867	6473	1.1	26273	36366	3363	4473	1.1
S. Korea	48	49	0.1	12792	30878	614	1513	3.5	14896	36000	715	1764	3.5
Australia/New Zealand	24	30	0.9	20875	35133	501	1054	2.9	28417	47767	682	1433	2.9
NON-OECD	5224	6903	1.1	1423	3823	7432	26393	5.0	4993	14464	26085	99848	5.3
Europe/Eurasia	342	319	−0.3	1912	6038	654	1926	4.2	9743	30975	3332	9881	4.3
Russia	144	125	−0.5	2285	6824	329	853	3.7	13243	39632	1907	4954	3.7
Other	198	193	−0.1	1641	5560	325	1073	4.7	7197	25534	1425	4928	4.9
Asia	3356	4231	0.9	1123	3704	3768	15672	5.6	4720	16417	15841	69460	5.8
China	1307	1446	0.4	1306	6053	707	8752	6.5	5908	27382	7722	39594	6.5
India	1087	1449	1.1	553	1738	601	2518	5.7	3425	10771	3723	15607	5.7
Other	962	1335	1.3	1517	3297	1459	4402	4.3	4567	10681	4393	14259	4.6
Middle East	191	301	1.8	3733	6847	713	2061	2.4	7607	14053	1453	4230	4.2
Africa	887	1463	1.9	794	1649	704	2412	4.8	2436	5064	2161	7408	4.9
Central and South America	448	589	1.1	3556	7336	1593	4321	3.9	7359	15058	3297	8869	3.9
Brazil	184	236	1.0	3565	6589	656	1555	3.4	7859	14530	1446	3429	3.4
Other	264	354	1.1	3549	7814	937	2766	4.2	7015	15367	1852	5440	4.2
TOTAL WORLD	6388	8203	1.0	5469	9369	34937	76850	3.1	8564	18812	54704	154313	4.1

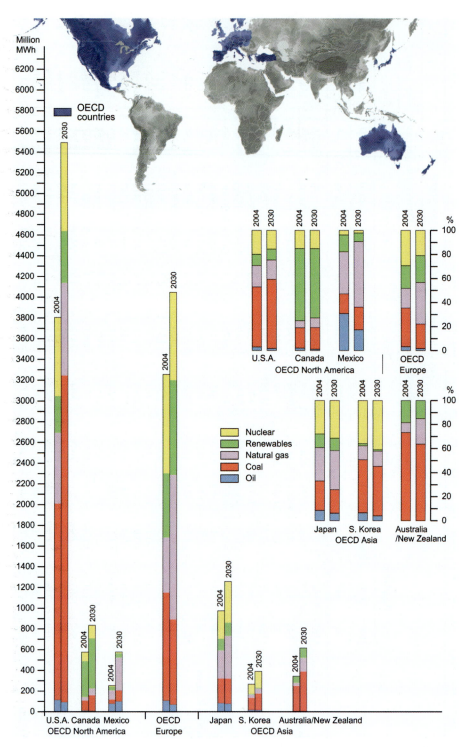

Figure 2.2.   OECD countries: Electricity generation and source mix in absolute and relative values for the year 2004 and forecasts for the year 2030 (data from EIA 2007).

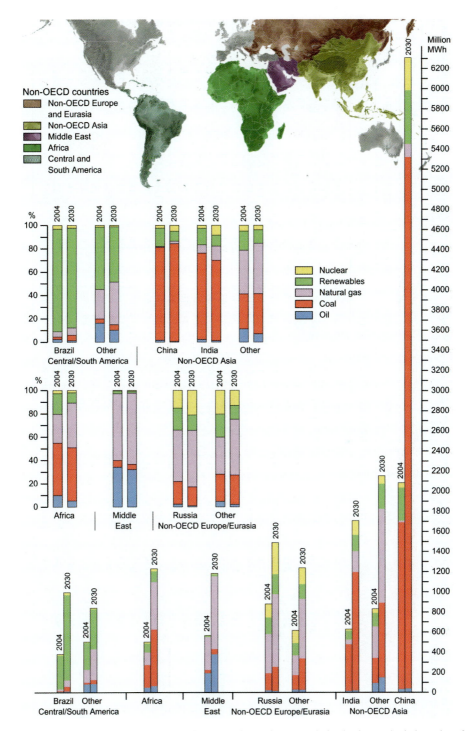

Figure 2.3. Non-OECD countries: Electricity generation and source mix in absolute and relative values for the year 2004 and forecasts for the year 2030 (data from EIA 2007).

Table 2.2. Total and by source electricity generation and respective source shares expressed in percent of total electricity generation for the year 2004 and projection for 2030 (data from EIA 2007).

	Total power generation				Coal					Oil				
	2004	2015	2030	Increase	2004		2030		Increase	2004		2030		Increase
	10^6 MWh	10^6 MWh	10^6 MWh	%/year	10^6 MWh	%	10^6 MWh	%	%/year	10^6 MWh	%	10^6 MWh	%	%/year
OECD	9626	11120	13500	1.3	3756	39.0	5159	38.2	1.2	416	4.3	386	2.9	−0.3
North America	4043	5579	7197	1.5	2117	52.4	3589	49.9	2.1	210	5.2	215	3.0	0.1
USA	3975	4516	5797	1.4	1979	49.8	3330	57.4	2.0	122	3.1	107	1.8	−0.5
Canada	573	708	836	1.5	98	17.1	152	18.2	1.7	12	2.1	10	1.2	−1.0
Mexico	243	352	564	3.3	40	16.5	107	19.0	3.9	75	30.9	98	17.4	1.0
Europe	3252	3564	4043	0.8	1052	32.3	826	20.4	−0.9	99	3.0	67	1.7	−1.5
Asia	1586	1976	2259	1.4	587	37.0	743	32.9	0.9	107	6.7	104	4.6	−0.1
Japan	977	1155	1257	1.0	240	24.6	245	19.5	0.1	83	8.5	78	6.2	−0.2
S. Korea	345	506	616	2.3	152	44.1	255	41.4	2.0	23	6.7	24	3.9	0.2
Australia/New Zealand	266	315	385	1.4	194	72.9	244	63.4	0.9	1	0.4	1	0.3	2.3
NON-OECD	6957	11169	17170	3.5	2966	42.6	8491	49.5	4.1	522	7.5	792	4.6	1.6
Europe/Eurasia	1496	2036	2731	2.3	316	21.1	554	20.3	2.2	49	3.3	41	1.5	−0.7
Russia	881	1121	1490	2.0	172	19.5	241	16.2	1.3	21	2.4	15	1.0	−1.2
Other	615	915	1238	2.7	144	23.4	313	25.3	3.0	29	4.7	26	2.1	−0.4
Asia	3518	6161	10186	4.2	2365	67.2	7229	71.0	4.4	138	3.9	207	2.0	1.6
China	2080	3728	6338	4.4	1658	79.7	5317	83.9	4.6	33	1.6	37	0.6	0.5
India	631	1098	1704	3.9	467	74.0	1174	68.9	3.6	14	2.2	20	1.2	1.5
Other	807	1335	2141	3.8	241	29.9	737	34.4	4.4	91	11.3	149	7.0	1.9
Middle East	566	849	1185	2.9	32	5.7	54	4.6	2.0	194	34.3	381	32.2	2.6
Africa	505	797	1235	3.5	227	45.0	566	45.8	3.6	50	9.9	65	5.3	1.0
Central and South America	883	1326	1838	2.9	26	2.9	87	4.7	4.8	91	10.3	98	5.3	0.3
Brazil	381	677	996	3.8	8	2.1	47	4.7	7.2	9	2.4	13	1.3	1.2
Other	501	649	843	2.0	18	3.6	40	4.7	3.2	82	16.4	86	10.2	0.2
TOTAL WORLD	16596	22289	30694	2.4	6723	40.5	13650	44.5	2.8	937	5.6	1198	3.9	0.9

(continued)

Table 2.2. *(continued)*

	Natural gas					Nuclear					Renewables				
	2004		2030		Increase	2004		2030		Increase	2004		2030		Increase
	10^6 MWh	%	10^6 MWh	%	%/year	10^6 MWh	%	10^6 MWh	%	%/year	10^6 MWh	%	10^6 MWh	%	%/year
OECD	1697	17.6	3282	24.3	2.6	2220	23.1	2526	18.7	0.5	1537	16.0	2147	15.9	1.3
North America	833	20.6	1317	18.3	1.8	883	21.8	1033	14.4	0.6	748	18.5	1043	14.5	1.3
USA	715	18.0	942	16.2	1.1	789	19.8	896	15.5	0.5	370	9.3	522	9.0	1.3
Canada	33	5.8	67	8.0	2.8	86	15.0	126	15.1	1.5	344	60.0	481	57.5	1.3
Mexico	85	35.0	307	54.4	5.1	9	3.7	11	2.0	0.9	34	14.0	41	7.3	0.7
Europe	531	16.3	1394	34.5	3.8	941	28.9	847	20.9	−0.4	629	19.3	909	22.5	1.4
Asia	333	21.0	571	25.3	2.1	396	25.0	646	28.6	1.9	163	10.3	195	8.6	0.7
Japan	271	27.7	411	32.7	1.6	274	28.0	394	31.3	1.4	109	11.2	129	10.3	0.7
S. Korea	40	11.6	78	12.7	2.6	124	35.9	252	40.9	2.8	6	1.7	7	1.1	0.7
Australia/New Zealand	22	8.3	82	21.3	5.1	0	0.0	0	0.0	–	49	18.4	58	15.1	0.7
NON-OECD	1522	21.9	4141	24.1	3.9	399	5.7	1093	6.4	4.0	1548	22.3	2653	15.5	2.1
Europe/Eurasia	579	38.7	1315	48.2	3.2	263	17.6	476	17.4	2.3	289	19.3	345	12.6	0.7
Russia	384	43.6	723	48.5	2.5	137	15.6	315	21.1	3.2	167	19.0	196	13.2	0.7
Other	195	31.7	592	47.8	4.4	125	20.3	161	13.0	1.0	122	19.8	146	11.8	0.7
Asia	360	10.2	1259	12.4	4.9	103	2.9	557	5.5	6.7	552	15.7	934	9.2	2.0
China	11	0.5	126	2.0	9.7	48	2.3	329	5.2	7.7	330	15.9	529	8.3	1.8
India	45	7.1	207	12.1	6.1	15	2.4	144	8.5	9.1	90	14.3	159	9.3	2.2
Other	303	37.5	926	43.3	4.4	40	5.0	84	3.9	2.9	132	16.4	245	11.4	2.4
Middle East	326	57.6	724	61.1	3.1	0	0.0	6	0.5	–	14	2.5	20	1.7	1.3
Africa	125	24.8	475	38.5	5.3	14	2.8	21	1.7	1.5	89	17.6	108	8.7	0.7
Central and South America	144	16.3	369	20.1	3.7	19	2.2	33	1.8	2.2	603	68.3	1251	68.1	2.8
Brazil	17	4.5	62	6.2	5.0	12	3.1	22	2.2	2.5	335	87.9	852	85.5	3.7
Other	126	25.1	307	36.4	3.5	7	1.4	11	1.3	1.6	268	53.5	399	47.3	1.5
TOTAL WORLD	3231	19.5	7423	24.2	3.3	2619	15.8	3619	11.8	1.3	3086	18.6	4804	15.7	1.7

to rank 3rd in the world. The entire non-OECD Asia region produced 21% of the world power in the year 2004 and is projected to increase its share to nearly 33% in the year 2030.

Middle East: In the Middle East the electricity generation is projected to increase on average by 2.9% per year, from 566 TWh in 2004 reaching 1185 TWh in 2030 (Fig. 2.3). This corresponds to 3.4% and 3.9% of the world power generation in 2004 and 2030, respectively.

Africa: The electricity generation in Africa is projected to increase by 3.5% per year i.e. from 505 TWh in 2004 to 1235 TWh in 2030. This is similar to the Middle East region (Fig. 2.3), representing 3.0 and 4.0% of the world power generation in 2004 and 2030, respectively.

Central and South America: In the Central and South America region, electricity generation is expected to increase from 883 TWh in 2004 to 1838 TWh in 2030, corresponding to an average increase of 2.9% per year (Fig. 2.3). This corresponds to 5.3 and 6.0% of the world power generation in the years 2004 and 2030, respectively. The strong and fast economic development of Brazil is reflected in the country's electricity generation, which increases from 381 TWh to 996 TWh in 2030 (corresponding to 54% of the regional power generation), that corresponds to an average increase of 3.8% per year, which is about double compared to the rest of the countries of the region (2.0% per year).

2.3   REGIONAL ELECTRICITY SOURCE MIX AND FORECASTS UNTIL 2030

The following forecasts of the energy source mix for electricity generation are based on the International Energy Outlook 2007 data base (IEO reference case, EIA 2007), which is largely influenced by the relatively high world market prices for oil and natural gas, which are expected to remain high during the 2004–2030 projection period. The presented forecasts are highly sensitive to the concerns of politicians and societies due to the environmental impact of greenhouse gas emissions from fossil fuel-based power plants, which result in an increasing interest in nuclear power (e.g. in India and China) and renewable energy sources (especially in OECD countries) as alternatives to fossil fuel power plants for electricity generation.

2.3.1   *Coal*

In the 2004–2030 world scenario (EIA 2007), coal remains the principal source for electricity generation, contributing from 41% in 2004 to 45% in 2030, to the world total electricity generation. Coal together with natural gas are the principal sources of electricity and have the highest projected annual growth rates compared to oil, nuclear, and renewables: 2.8% for coal (OECD: 1.2%; non-OECD: 4.1%), and 3.3% for gas (OECD: 2.6%; non-OECD: 3.9%) so that the coal/gas relation remains more or less constant (about 2:1) during 2004–2030 period.

As long as oil and natural gas prices are high, coal continues to be favored for power generation, especially by countries with rich coal resources like China, India and the USA, where coal was used in 2004 to generate 80, 74, and 50% of the national electricity, respectively. These coal shares are expected to further increase until 2030 in China (annual increase of coal use: 4.6% per year) and USA (annual increase of coal use: 2.0% per year) reaching a share of 84% in China and 57% in USA for national power generation, whereas the coal share in India will decrease to 69% (although the annual average increase of coal use is 3.6%) of the national power sources. Other important regions whose principal electricity source is coal are Australia/New Zealand (2004: 73%; 2030: 63%), and Africa (2004: 45%; 2030: 46%). In OECD Europe, coal share as a source for power generation is projected to decrease significantly in the 2004–2030 period from 32% in 2004 to 20% in 2030. Worldwide, OECD Europe is the only region where the average annual increase rate of coal use is negative (–0.9% per year). The same trend can be observed in Japan, where the share of coal is expected to decrease from 25 to 20% in the 2004–2030 period. In other regions, where coal is a significant source for electricity generation, slight or moderate changes can be expected for the 2004–2030 period: Canada: a slight increase from 17 to 18%, Mexico: an increase from 17 to 19%, South Korea: a decrease from 44 to 41%, non-OECD Asia (excluding China and India): an increase from 30 to 34%, Russia: a decrease from 20 to 16%, and non-OECD Europe/Eurasia (without Russia): a slight increase from 23 to 25%. Other countries or regions where the share of coal for electricity generation is below 10% are the Middle East and the region of Central and South America.

2.3.2 *Natural gas*

Natural gas is the second most important source contributing to worldwide electricity generation, with 20% in 2004 and a projected 24% in 2030. As already mentioned in Chapter 2.3.1, gas has the highest increase rate of all sources used for power generation.

In all the countries considered by IEO 2007 where natural gas was the principal electricity source in 2004 the share of this source is expected to further increase moderately or strongly in the 2004–2030 period: Mexico from 35 to 54% (5.1% per year); Japan from 28 to 33% (1.6% per year); Russia from 44 to 49% (2.5%/year); non-OEDC Europe/Eurasia (without Russia) from 32 to 48% (4.4% per year); non-OECD Asia (without China and India) from 38 to 43% (4.4% per year); Middle East from 58 to 61% (3.1% per year); and Central/South America (without Brazil) from 25 to 36% (3.5% per year). If we look at OECD Europe, whose principal electricity source was coal in the year 2004 (32%), the gas share was only 16%. We can see that this coal/gas ratio becomes inverted over time, until 2030 when natural gas, which is expected to have an average annual growth rate of 3.8% will account for 35% of the regional electricity production, whereas the share of coal is projected to decline to 20%. This development can be seen in the framework of OECD Europe's strong attempts to reduce its greenhouse gas emissions. Also significant increases of the natural gas share for electricity production are predicted for Africa, where the natural gas share is expected to increase from 25% in 2004 to 39% in 2030 (average by 5.3% per year) and for Australia/New Zealand, where the natural gas share is forecasted to increase in the same period from 8 to 21% (in average by 5.1% per year). Also, USA has a significant natural gas share; however it is expected to reduce from 18% in 2004 to 16% in 2030. In all these three aforementioned regions, coal will remain the primary source for electricity generation.

2.3.3 *Oil*

Oil contribution to world electricity generation was only 5.6% in 2004 and is expected to fall to 3.9% in 2030. This source has the lowest increase rate since oil prices are expected to remain high (projected to be US$ 59 per barrel in 2030; given in 2005 US$). Hence, its use for power generation is projected to increase worldwide only on an average of 0.9% per year, with an increase of 1.6% per year in the non-OECD countries and an increase of −0.3% in the OECD countries. These low average increase rates are due to a decrease in oil in the source mix for electricity generation. Only the Middle East and Mexico have a considerable oil-share in their power source mix. In the Middle East, this share is projected to remain more or less constant during the 2004–2030 period (2004: 34%; 2030: 32%), whereas in Mexico oil is increasingly substituted by natural gas and its share is expected to drop sharply from 31 to 17% in this period. In all other regions and individual large countries considered in the IEO-projection, the share of oil as source for power generation is projected to account for less than 10% in 2030. However, it needs to be noted that the high oil-dependence of several individual countries is not reflected in this projection, as we shall see in Chapter 4.3.2.

2.3.4 *Nuclear*

The worldwide average increase of nuclear energy for power generation is projected to be 1.3%, which implies that its share in total electricity production could decrease from 16% in the year 2004 to 12% in 2030. However, significant differences between the OECD region with an increase of only 0.5 and a 4.0% increase in the non-OECD region indicates that the cause is mainly due to the nuclear program of China and India, resulting in projected forecasts of an average growth rate of 7.7% per year in China and 9.1% per year in India for the period 2004–2030. Thus, China is projected to increase the share of nuclear power in its mix for national electricity generation from 2.3 to 5.2%, and the corresponding increase in India is expected to be from 2.4 to 8.5%, In the countries that are already using a significant percentage of nuclear share in the electricity source mix, the share is expected to decrease in the 2004–2030 period: in USA from 20 to 16%, in OECD Europe from 29 to 21%, and in non-OECD Europe/Eurasia (without Russia) from 20 to 13%, whereas in Canada the share remains constant at 15%, while in Japan it increases from 28 to 31%, in South Korea from 36 to 41%, and in Russia from 16 to 21%. In the other regions considered in this outlook, the nuclear share for power generation is less than 5%.

2.3.5   *Renewables*

Even if renewable energy resources, considered as zero emission sources, are attractive since oil and natural gas prices are expected to remain high, renewables are projected to increase worldwide only on average by 1.7% per year during the 2004–2030 period (OECD: 1.3%; non-OECD: 2.1%). As a consequence, the share of renewable energy resources for world electricity generation is expected to decline from 19% in 2004 to 16% in 2030. The renewables used for power generation correspond predominantly to hydroelectric power. In OECD Europe wind energy's share for power generation is considerable, whereas high-temperature geothermal resources are not significant on a worldwide scale, even though its share in some selected countries is significant in generating electricity (e.g. in the year 2005 in decreasing order of percentage: El Salvador 22%, Kenya 19%, Philippines 19%, Iceland 17%, Costa Rica 15%, Nicaragua 10%, Indonesia 7%; for details see Chapter 3.4.1). Although low-enthalpy geothermal resources have recently started to be promoted and used in some OECD countries for power generation, they do not yet contribute significantly to the national power production; non-OECD countries are, with exception of Nagqu plant in China, not yet using low-temperature resources for commercial power generation.

Canada and Brazil are the countries that produce the largest amount of electricity using renewables, mainly hydroelectric power. In the year 2004 Canada generated 60% and Brazil generated 88% of its national electricity through hydroelectric power. These values are expected to decrease until 2030 slightly to 58 and 86%, respectively.

Thus, Brazil's dependence on hydroelectric power is most suitable for low greenhouse gas emission, but over dependence on hydroelectric power may result in power shortage since this source is sensitive to climatic events like rainfall and droughts. Thus, this situation calls for source diversification, especially for countries like Brazil (this issue will be discussed in more detail in Chapter 4.4). In the remaining Central/South American region (without Brazil) the use of renewables, predominantly hydroelectric power, is also high, amounting to 54% in 2004 and is expected to decrease to 47% by 2030.

Among the other regions, where renewables had a share of over 10% in the source mix used for power generation in 2004, OECD Europe is the only region where this share is expected to increase during the projected period from 19 to 23%, which is related to strong promotion of renewable energy sources drastically reducing greenhouse gas emissions. Since in OECD Europe, most of the economically feasible hydroelectric resources have already been developed, many countries are setting new goals to increase the use of other non-hydroelectric power renewables such as wind and geothermal energy. Especially during the last few years, schemes to generate power from wind turbines have been implemented, and at the end of 2006 the 27-member European Union accounted for 65% of the world total installed generation capacity through wind turbines (AWEA 2007).

In other regions where in 2004 renewables had a share of more than 10%, these shares are expected to decline heavily in the future: i.e Mexico (which had in 2005 3% geothermal power) from 14 to 7%; Africa from 18 to 9%; China from 16 to 8%; India from 14 to 9%; non-OECD Asia (without China and India) from 16 to 11%; Russia from 19 to 13%; non-OECD Europe/Eurasia (without Russia) from 20 to 12%; and Australia/New Zealand from 18 to 15%. In Japan the renewables share is expected to decline slightly from 11 to 10%; in USA however, it remains constant, contributing to 9% (including 0.5% of geothermal power in 2004) of the national electricity generation.

CHAPTER 3

Worldwide potential of low-enthalpy geothermal resources

"Earth's currently and potentially available reserve of geothermal energy is a quantity of astonishing magnitude—vastly greater, in fact, than the resource bases of coal, oil, gas, and nuclear energy combined. . . . Although only a fraction of this geothermal bounty can now be tapped, with innovative technology it will remain available for our descendants long after the last drop of oil is produced."

University of Utah: Geothermal Energy, 2001.

3.1 WORLD GEOTHERMAL RESOURCES

The geothermal resources of the earth are huge. The part of geothermal energy stored at a depth of 3 km is estimated to be 43,000,000 EJ corresponding to 1,194,444,444 TWh (Bijörnsson *et al.* 1998). Even this small part (<1%) that corresponds to the currently available share, and can be extracted economically using existing technology, is vast compared to the total world net electricity generation, which is expected to grow from 16,424 TWh in 2004 by 85% to reach 30,364 TWh in the year 2030 (EIA 2007). Geothermal resources are also much larger compared to all fossil fuel resources put together, whose energy and electricity equivalents are 36,373 EJ and 1,000,400 TWh, respectively corresponding to (1) oil: 1317.4 billion barrels (corresponding to 8062 EJ or 223,900 TWh), (2) natural gas 6183 trillion cubic feet (corresponding to 6678 EJ or 185,500 TWh), and (3) coal: 998 billion short tons (corresponding to 21,634 EJ or 600,900 TWh) (EIA 2007). If we consider that only a small part of the geothermal energy resources can be tapped for example, from either low-enthalpy wet geothermal systems (WGS), or enhanced geothermal systems (EGS) applying artificial fracturing of the geothermal reservoir in both cases using advanced heat exchanger technologies which reduce the minimum fluid temperature required for power generation, the world geothermal resource will remain available for future generations long after the last drop of oil is exploited. Continuous development of innovative drilling and power generation technologies makes this source the best future option available to meet the required future electricity demand of the world, drastically reducing greenhouse gas emissions and mitigating global climate change.

The potential of low-enthalpy geothermal resources (<150 °C), which are suitable for electricity generation (at present thermal waters with temperature of 74 °C are being utilized for power generation) is much higher than that of high-enthalpy resources (>150 °C) since these reservoirs are widespread and occur at shallow depths. These reservoirs include both shallow wet geothermal systems that surround the high-enthalpy systems, and also the low-temperature conduction dominated enhanced geothermal systems. (Fig. 3.1)

Low-enthalpy geothermal resources in developing countries are yet to be exploited extensively for electricity generation, although, they are already being utilized presently, as discussed in Chapter 10, in industrialized countries for generating electricity. Low-enthalpy geothermal resources are being utilized for direct applications in several developing countries in spite of their high potential to generate electricity. Even the high-enthalpy resources in developing countries, as discussed in Chapters 3.3 and 3.4, are not exploited to the extent that they could be.

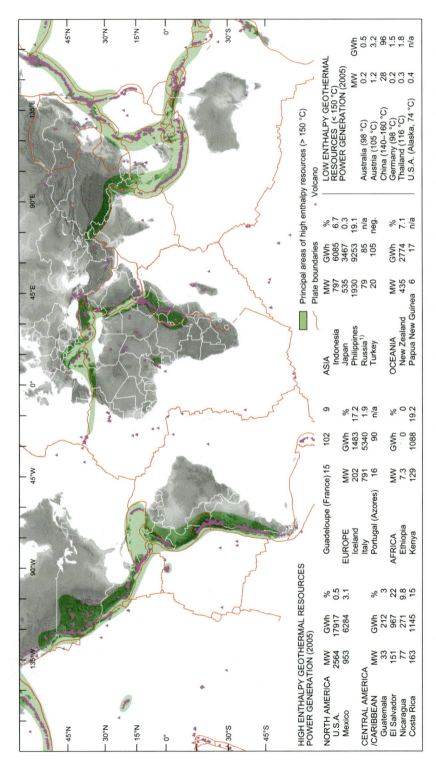

HIGH ENTHALPY GEOTHERMAL RESOURCES POWER GENERATION (2005)

	MW	GWh	%
NORTH AMERICA			
U.S.A.	2564	17917	0.5
Mexico	953	6284	3.1
	MW	GWh	%
CENTRAL AMERICA			
/CARIBBEAN			
Guatemala	33	212	3
El Salvador	151	967	22
Nicaragua	77	271	9.8
Costa Rica	163	1145	15

	MW	GWh	%
Guadeloupe (France)	15	102	9
EUROPE	MW	GWh	%
Iceland	202	1483	17.2
Italy	791	5340	1.9
Portugal (Azores)	16	90	n/a
AFRICA	MW	GWh	%
Ethiopia	7.3	0	0
Kenya	129	1088	19.2

	MW	GWh	%
ASIA			
Indonesia	797	6085	6.7
Japan	535	3467	0.3
Philippines	1930	9253	19.1
Russia[1]	79	85	n/a
Turkey	20	105	neg.
OCEANIA	MW	GWh	%
New Zealand	435	2774	7.1
Papua New Guinea	6	17	n/a

LOW ENTHALPY GEOTHERMAL RESOURCES (< 150 °C) POWER GENERATION (2005)

	MW	GWh
Australia (98 °C)	0.2	0.5
Austria (105 °C)	1.2	3.2
China (140–160 °C)	28	96
Germany (98 °C)	0.2	1.5
Thailand (116 °C)	0.3	1.8
U.S.A. (Alaska, 74 °C)	0.4	n/a

Principal areas of high enthalpy resources (> 150 °C)
Plate boundaries
Principal areas of high enthalpy resources (> 150 °C)
Volcano

Figure 3.1.   Distribution of high- and low-enthalpy geothermal resources along plate boundaries and in active volcanic zones (schematic) and classification of countries with commercial geothermal power generation with installed geothermal capacity (MW), geothermal electricity generation (GWh) and share of total national electricity production in the year 2005 (data from Bertani 2005).

3.2   TYPES OF GEOTHERMAL SYSTEMS

Geothermal systems can be classified based on their association with the tectonic regime in different geological domains, described in detail in Chapter 5. Most of them are cyclic systems with rain water as the main carrier of heat from the deeper parts of the earth to the surface. Depending on the local geological and thermal regimes, the systems could be steam-dominated or liquid-dominated systems. Depending on the temperature of the reservoirs, these systems evolve into low- ($<150\,°C$) and high-enthalpy ($>150\,°C$) resources. To be viable for exploitation, these systems should be accessible at reasonable depths with sufficient geothermal fluids to sustain long productivity. High-temperature hydrothermal systems, used in the year 2007 in 19 countries for power generation, are restricted to plate boundaries and areas of active volcanism (Fig. 3.1), while low-enthalpy resources are available in a variety of geological and tectonic settings. Sometimes they occur close to high-enthalpy resources and at times they occur independently as large exploitable resources. Low-enthalpy geothermal resources occur also as geopressured systems in large sedimentary basins (see Chapter 5) that have not been exploited for commercial applications. Estimates of the world low- and high-enthalpy potentials are detailed in section 3.3, and section 3.4 gives an overview of the resources for commercial production of electric power. High- and low-temperature conduction dominated enhanced geothermal systems are gaining importance at present and may provide a viable solution to the electricity demand facing all countries (MIT 2006). With the advancement made in heat exchanger and drilling technologies, EGS may be able to provide low-enthalpy fluids at shallow depths in all countries. According to the MIT report on EGS (MIT 2006), the USA alone has a potential of about 13,000,000 EJ within a depth range of 3 to 10 km. From this potential, 200,000 EJ can be extracted for utilization—corresponding to about 2000 times the annual consumption of primary energy by the USA in 2005 (MIT 2006), and by 2050 the USA should be in a position to economically generate about 100,000 MW_e with modest R and D investment.

This fact should attract the attention of all the governments and energy policy makers among countries where geothermal energy has yet to be considered as a viable alternate source of meeting future power demands. They should consider low-enthalpy geothermal resources as well as EGS in the middle to long-term planning of energy policies of their countries, and they should take immediate initiatives to evaluate such resources.

3.3   AVAILABLE LOW- AND HIGH-ENTHALPY GEOTHERMAL RESOURCES

In most developing countries, the geothermal potential is not well known, and only estimations are available. The potentials of low-enthalpy resources especially, did not receive attention, not only in most of the developing countries, but also in several industrialized countries. The main reason for not developing these resources for commercial exploitation is that they are not considered as economically viable for electricity generation. Although this has been true in the past, technological advances made in the field of heat exchangers and drilling methods in the last decade allow geothermal fluids with temperatures as low as $74\,°C$ to be used for electric power generation (see Chapter 9). These developments urgently call for all countries to evaluate low-enthalpy resources, to include these resources in national planning and exploit these resources immediately, thereby reducing greenhouse gas emissions and improving the quality of the environment, especially in developing countries.

A worldwide country-by-country assessment of geothermal potential was published in 1999 by the Geothermal Energy Association (GEA) in Washington D.C. (Gawell *et al.* 1999) and is shown in Table 3.1. Worldwide, the known geothermal potential is of the order of 1089 TWh/year, and 73% of this is found in Latin America, Asia and Pacific, and Africa (Gawell *et al.* 1999).

Bijörnsson *et al.* (1998) developed a method that uses the distribution of active volcanoes and their correlation to the geothermal activity to assess the high-temperature geothermal resources. The number of active volcanoes in a specific country is correlated to the country's estimate of its

Table 3.1.   Regional geothermal energy potentials.

Region	Gawell et al. 1999[1] Known geothermal potential TWh/yr	Bijörnsson et al. 1998 Useful accessible resource base Electricity[3] TWh/yr	Direct use EJ	Stefansson 1998 High-temperature; electricity production TWh/year Convent. technology	Convent. and binary technology[5]	Low-temperature; direct use[4] EJ/yr
North America	200	1482	75555	1330	2700	>120
Latin America	354	3112	100969	2800	5600	>240
Europe	97	2030[2]	105035	1830	3700	>370
Asia and Pacific	337	4465	170007	4020	8000	>430
Africa	101	1354	146936	1220	2400	>240
World	1089	12443	598529	11200	22400	>1400

[1] Based on advanced technology potential data from Gawell et al. 1999; [2] Inclusive CIS and Turkey; [3] Assumed capacity for a sustainable generation; [4] Lower limit; [5] >100 °C.

electricity generation from UARB (Useful Accessible Resource Base; less than 3 km deep), which corresponds to the identified and not yet identified resources, which are economically accessible. Performing this correlation for the countries of the USA, Iceland, Italy, Indonesia, Philippines, Japan, New Zealand, and Mexico, a correlation factor of 9.4 was obtained

$$\text{producible electricity from UARB (TWh/yr)} = 9.4 \times \text{number of active volcanoes} \qquad (1)$$

Using this correlation coefficient for all the regions of the world, the authors obtained the regional high-temperature resources potentials as shown in Table 3.1 resulting in a world potential of about 12,000 TWh/year.

To assess the low-temperature potential of different regions of the world, Bijörnsson et al. (1998), used the existing estimates of low-temperature resources in Hungary and Poland. They used a correlation factor between the UARB and the ARB (Accessible Resource Base) for these regions, which is reported by Aldrich et al. (1978) as 0.0114. These authors assume that the correlation factor obtained for these countries can be used for all regions of the world to estimate the low-temperature resources. The estimated regional UARB values given in Table 3.1 indicate that the low-enthalpy geothermal resource potential of the world is about 600,000 EJ for direct use that corresponds to a power generation of about 165,600,000 TWh.

A study by Stefansson (1998) estimated "identified, suspected and not yet identified" geothermal resources (Tab. 3.1) according to their suitability for electricity generation using either conventional or binary fluid technology (his minimum temperature criteria was 100 °C). His estimate indicates that the use of geothermal fluids with at least 100 °C will double the energy available for electricity generation from 11,200 TWh/year using conventional technology to 22.400 TWh/year using both conventional and binary technologies (Table 3.1). This study is based on the distribution of active volcanoes to determine the high-temperature resources. It uses an empirical relation between the frequencies of high- to low-temperature resources to determine the low-temperature resources potential. According to the authors, the ratio between "total" and "identified" resources is about 5 to 10. This indicates that the Geothermal Energy Association (GEA) estimates are lower and that in reality the geothermal reserves are, depending on the region, on average 5 to 10 times higher than that reported by GEA.

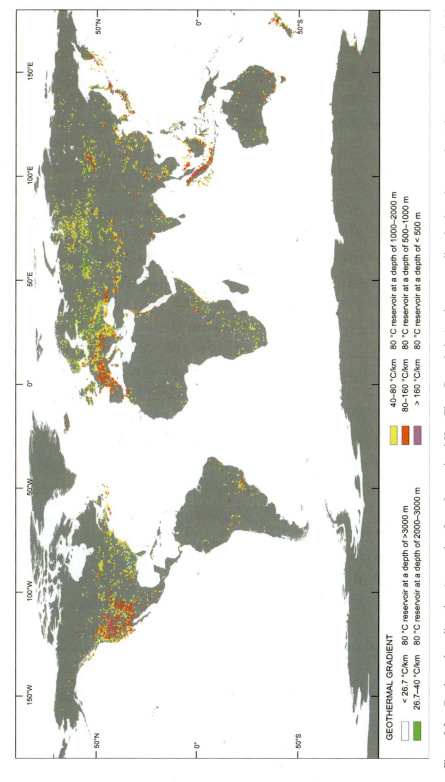

GEOTHERMAL GRADIENT

< 26.7 °C/km	80 °C reservoir at a depth of >3000 m
26.7–40 °C/km	80 °C reservoir at a depth of 2000–3000 m
40–80 °C/km	80 °C reservoir at a depth of 1000–2000 m
80–160 °C/km	80 °C reservoir at a depth of 500–1000 m
> 160 °C/km	80 °C reservoir at a depth of < 500 m

Figure 3.2.   Geothermal gradient data (source data from: International Heat Flow Commission) and corresponding depths, where geothermal resources with a suitable temperature for power generation of minimal 80 °C can be expected.

A comparison of the estimates of Stefansson (1998) and Bijörnsson *et al.* (1998) shows good agreement with respect to high-temperature resources (12,443 *versus* 11,200 TWh/yr), whereas the differences with respect to low-temperature estimates are highly varying by a factor of ten according to Stefansson (1998), who indicates that the method gives only a lower limit for the low-temperature resources. Based on these estimates, it is certain that the potential of the geothermal energy resources, especially the low-enthalpy resources, is vast. However this potential of low-enthalpy geothermal resources for power generation has not been realized by developing countries and in certain countries it is practically untapped.

The evaluation, considering binary technologies, both by Gawell *et al.* (1999) and Stefansson (1998) showed that the geothermal potential will be doubled if low-temperature resources of temperatures between 100 and 150 °C are also considered for power generation. This apperars to be inappropriate since their estimates considered only the regions with high-enthalpy systems and did not include other regions with potential low-enthalpy resources. It appears that the low-enthalpy resources volume is manyfold higher than the high-enthalpy resources. Hence, a standardized assessment method should be developed to estimate the low-enthalpy resources for electricity generation for the entire world. One way of making an assessment of accessible low-enthalpy resources potential is by looking at the geothermal gradient map (Fig. 3.2). This map will give an estimation about the depth of 80 °C isotherms that can be tapped to generate power using the present day available advanced technology. The only drawback in this map is that it does not cover several regions in developing countries. However, it is possible to update regional heat flow maps on the basis of several abandoned wells drilled for locating high-temperature resources in these regions. These wells were abandoned in the past since low-enthalpy resources potential had not been realized yet. This increase in the geothermal potential for power generation due to the inclusion of unutilized low-enthalpy geothermal resources (Bundschuh and Chandrasekharam 2002) that can be accessed to meet the growing energy demand, and to reduce the increasing CO_2 emission in the atmosphere, makes it necessary to target the exploration and exploitation of low-enthalpy resources in the future by all countries.

3.4   ACTUAL USE AND DEVELOPMENTS OF LOW- AND HIGH-ENTHALPY GEOTHERMAL RESOURCES FOR POWER GENERATION

Geothermal energy has been used for power generation for the past century. However, worldwide, the geothermal option for national power generation has been ignored or underemphasized by many countries. Reasons are manifold. In the past, countries like the USA considered its high-temperature hydrothermal resources and evaluated them as insignificant and locally limited to be of importance to cover an important source share of the national power generation. As a consequence, funding for R&D was low and governmental actions, like state incentives to promote the geothermal development, were not applied. In the cases of many developing countries available geothermal resources were not evaluated or under emphasized, mostly due to regulatory, economic, and financial barriers. Some principal barriers, and ways and needs to overcome them are discussed in section 3.5.

This worldwide under-estimation of geothermal energy as a long-term option to provide base-load power, which in comparison with other low-emission options such as wind or solar energy does not require electricity storage systems, has led to a slowing down of the development of appropriate technologies to use geothermal resources as a viable option to meet the future electricity demand.

In developing countries, low-enthalpy geothermal resources especially, have not received much consideration for electricity generation. In contrast, in the industrialized countries, especially Europe and recently the USA, increasing energy demand and environmental awareness related to climate change have compelled these countries to develop technologies which use low-enthalpy geothermal resources economically for power generation. The installation of several commercial low-temperature geothermal plants like those at Hysavik in Iceland, Chena in Alsaka, Altheim in

Austria, and Neustadt-Glewe in Germany (see Chapters 9 and 10), have substantially proved the ability of low-enthalpy geothermal fluids to generate electricity.

These developments are coherent with the international progress in the development of EGS (low- and high-temperature conduction dominated EGS) and related technologies which have received much interest in Europe in the last decade. Installed capital costs for surface conversion plants are less for high-enthalpy resources (US$ 1500/kW for 400 °C resource temperature compared to US$ 2300/kW for 100 °C resource temperature; MIT 2006), but require higher drilling costs compared to low-enthalpy resources. For example, in the case of the Soultz project, which is predominantly financed by the European Union, by artificially fracturing an active geothermal reservoir a volume of over 2 km^3 has been created at depths between 4 and 5 km with fluid production rates exceeding by a factor of 3 relative to the initial commercial goals (MIT 2006).

Recent increases in the cost and uncertainty of future conventional energy supplies are improving the attractiveness of low-enthalpy geothermal resources and EGS in general. In 2006 it received much attention in the USA, since this country wants to secure its electricity supply from domestic resources to reduce its dependence on oil and gas imports, and at the same time reduce it's contribution to global warming. Due to mounting international pressure, the USA, which did not sign the Kyoto protocol, has highlighted in the June 2007 G8 meeting in Heiligendamm (Germany) it's wishes to participate in a post-Kyoto agreement to reduce greenhouse gas emissions by 50% until the year 2050. These reasons cause a growing awareness of the genuine value and near limitless potential of these practically unused geothermal resources. In this context, it is significant to see the huge geothermal potential of the USA, especially of low-temperature/EGS resources which can be economically and technically exploited to cover the entire electricity demand of the USA. If this change occurs, EGS can become a major electricity source for base-load power generation in the USA by the year 2050.

Developing countries need to benefit from these new and continually improving technologies for using low-temperature geothermal resources. These resources may be destined to become a major factor in covering the increasing electricity demand in the developing world through using domestic resources while contributing to a reduction in global greenhouse gas emissions. However, most developing countries (as well as industrialized countries!) have not yet estimated their low-temperature geothermal potential. Countries which have high-enthalpy geothermal resources have focused on the estimation of their potential, whereas other countries which do not count on high-enthalpy resources, have not yet considered using the low-enthalpy resources for electricity generation.

Developing countries can access all available low-enthalpy wet and EGS sources for electricity generation immediately. For many developing countries, the use of low-enthalpy resources is not new. Many of them are aware (71 countries are reported with production data; Lund *et al.* 2005) (Fig. 3.3) and have been using these resources for the past centuries for bathing, and for the last few decades, for direct use, especially for space heating, bathing, domestic heat pumps, snow melting, heating greenhouses, aquaculture, drying of fruits, etc. Now, the challenge for these countries is to properly evaluate and to use these low-enthalpy resources for electricity generation.

3.4.1 *Countries with experiences using high-enthalpy resources for power generation*

The countries which are already using high-temperature reservoirs for power generation need to be convinced to focus not only on their isolated high-temperature reservoirs, but also on their low-temperature resources which cover much larger parts of their countries than high-enthalpy resources do. The countries which have exploited their high-temperature resources and found them unsuitable for power generation need to reassess and consider their low-temperature resources. This requires countrywide exploration of the geothermal potential followed by pre-feasibility studies for setting up power plants.

At present there are only 16 countries that are using high-enthalpy geothermal resources for commercial power generation (Fig. 3.1). However, there are several additional countries that had earlier

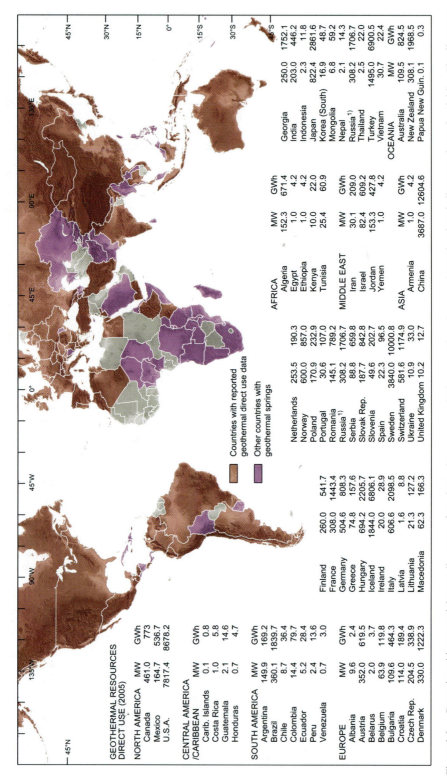

Figure 3.3.   Countries with reported data on direct use of geothermal resources and other countries where geothermal springs indicate geothermal activities. The direct use data correspond to the year 2005 and are complied from Lund *et al.* 2005. [1] Includes European and Asian parts.

GEOTHERMAL RESOURCES DIRECT USE (2005)

NORTH AMERICA	MW	GWh
Canada	461.0	773
Mexico	164.7	536.7
U.S.A.	7817.4	8678.2

CENTRAL AMERICA /CARIBBEAN	MW	GWh
Carib. Islands	0.1	0.8
Costa Rica	1.0	5.8
Guatemala	2.1	14.6
Honduras	0.7	4.7

SOUTH AMERICA	MW	GWh
Argentina	149.9	169.2
Brazil	360.1	1839.7
Chile	8.7	36.4
Colombia	14.4	79.7
Ecuador	5.2	28.4
Peru	2.4	13.6
Venezuela	0.7	3.0

EUROPE	MW	GWh
Albania	9.6	2.4
Austria	352.0	619.5
Belarus	2.0	3.7
Belgium	63.9	119.8
Bulgaria	109.6	464.3
Croatia	114.0	189.4
Czech Rep.	204.5	338.9
Denmark	330.0	1222.3
Finland	260.0	541.7
France	308.0	1443.4
Germany	504.6	808.3
Greece	74.8	157.6
Hungary	694.2	2205.7
Iceland	1844.0	6806.1
Ireland	20.0	28.9
Italy	606.6	2098.5
Latvia	1.6	8.8
Lithuania	21.3	127.2
Macedonia	62.3	166.3
Netherlands	253.5	190.3
Norway	600.0	857.0
Poland	170.9	232.9
Portugal	30.6	107.0
Romania	145.1	789.2
Russia [1]	308.2	1706.7
Serbia	88.8	659.8
Slovak Rep.	187.7	842.8
Slovenia	49.6	202.7
Spain	22.3	96.5
Sweden	3840.0	10000.8
Switzerland	581.6	1174.9
Ukraine	10.9	33.0
United Kingdom	10.2	12.7

AFRICA	MW	GWh
Algeria	152.3	671.4
Egypt	1.0	4.2
Ethiopia	1.0	4.2
Kenya	10.0	22.0
Tunisia	25.4	60.9

MIDDLE EAST	MW	GWh
Iran	30.1	209.0
Israel	82.4	609.2
Jordan	153.3	427.8
Yemen	1.0	4.2

ASIA	MW	GWh
Armenia	1.0	4.2
China	3687.0	12604.6
Georgia	250.0	1752.1
India	203.0	446.2
Indonesia	2.3	11.8
Japan	822.4	2861.6
Korea (South)	16.9	48.7
Mongolia	6.8	59.2
Nepal	2.1	14.3
Russia [1]	308.2	1706.7
Thailand	2.5	22.0
Turkey	1495.0	6900.5
Vietnam	30.7	22.4

OCEANIA	MW	GWh
Australia	109.5	824.5
New Zealand	308.1	1968.5
Papua New Guin.	0.1	0.3

Countries with reported geothermal direct use data

Other countries with geothermal springs

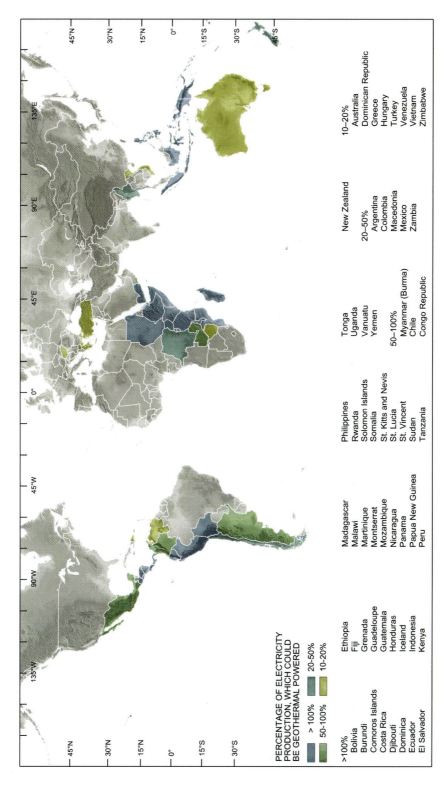

Figure 3.4.   Country-by-country percentage of electricity production, which could be generated from geothermal resources as established by a study of Gawell *et al.* (1999).

operated geothermal power plants and completed exploration work on high-enthalpy resources. These countries have the basic knowledge to assess the low-enthalpy resources potential of their country. Exploration studies carried out by several countries in the past, located geothermal reservoirs that were not considered economical at that time, but with the available advanced drilling and heat exchanger technology these countries should be encouraged to reconsider developing low-enthalpy resources for power generation.

As already mentioned, the high-enthalpy geothermal resources are located in regions along the plate boundaries and in young volcanic regions. Although such resources are found in many developing countries (Fig. 3.1), by far, not all of the countries are using these resources. Only a very few of these resources have been exploited, or even been considered for use. Figure 3.1 gives an overview on these countries including their installed geothermal capacity, the produced electricity and the percentage of their national power produced from high-enthalpy resources.

Figure 3.4, prepared based on the study by Gawell *et al.* (1999) shows the countries, and the percentage of total national electricity production, which could theoretically be produced geothermally using enhanced technology (wet geothermal systems with fluid temperature $>100\,^{\circ}C$). There are 39 countries, mostly from Latin America, the Caribbean, Eastern Africa and the Pacific region that can be theoretically exclusively geothermally powered.

If we compare the geographic location of the above mentioned countries with the countries where geothermal energy is used for power generation (Fig. 3.1), one realizes that at present, none of the South American countries located in the high-enthalpy Andean mountain chain uses geothermal power generation. Most of these countries use domestic fossil fuel resources. In contrast, Central America, which counts on about the same quantity of geothermal resources (advanced technology potential: 13,210 MW; Gawell *et al.* 1999) as South America (14,660 MW), but does not have significant fossil fuel resources (with exception of Guatemala, which has some oil), uses geothermal energy significantly for national electricity generation (Fig. 3.1): El Salvador 22%, Costa Rica 15%, Nicaragua 10%, and Guatemala 3% (2005). In North America (which includes Mexico) electricity produced from geothermal resources accounts for 0.5% in USA and 3.1% in Mexico of national power for the year 2005 (Bertani 2005) making USA as the world-wide highest absolute power generator from geothermal sources (17,917 GWh).

If we look at the many small islands in the Caribbean (e.g., Nevis, St. Kitts, Montserrat, Guadeloupe, Dominica, Martinique, St. Lucia, St. Vincent), the Indian Ocean (e.g., Comoros islands), and the Pacific (e.g., Tonga, Vanuatu), these islands could be, according to the analysis of Gawell *et al.* (1999), completely geothermally powered. As of today though, none of these islands is using locally available geothermal resources and instead depend on imported oil and gas for electricity generation. Only Guadeloupe (belonging to France) uses geothermal source for electricity generation amounting to 2% of total electricity produced in the island. The benefits of using geothermal resources for generating electricity by these small islands are discussed in Chapter 4 (section 4.3.1).

Compared to North and South America and the Pacific region, the high-enthalpy resources potential is the lowest in Africa region. However, it must be considered that Africa is also the region with the lowest power demand. The high-temperature geothermal resources of Africa are restricted to the rift valleys as described in Chapter 5. All the high-enthalpy geothermal provinces are surrounded by low-enthalpy regions and all these countries could be completely geothermally powered. In Africa, only Kenya generates about 19% (2005) of the total national electric power from geothermal sources. Other countries like Yemen, Mozambique and Madagascar also fall in the above category.

In the case of Asia and the Pacific islands, large high-enthalpy geothermal resources are available. Indonesia and Philippines are the two countries producing electricity from geothermal resources. These two countries have large potential (Indonesia 15,630 MW_e and Philippines 8620 MW_e, Gawell *et al.* 1999). These countries' share in the national power generation is 6.7 and 19.1% respectively. Both the countries have the potential to meet their electricity demand from geothermal sources alone. Papua New Guinea has recently started commercially exploiting its high-enthalpy geothermal sources. In 2005 it has generated 7.1% of total electricity from geothermal source.

3.5 OVERCOMING BARRIERS TO GEOTHERMAL ENERGY

Worldwide, low- and high-enthalpy geothermal resources, suitable for power generation are "under utilized market opportunities" and hence call for urgent development. This lack of development is due to regulatory, institutional, economic and financial barriers. These barriers exist in spite of the fact that low-enthalpy resources are available through out the year and have large social, environmental and economic benefits for all the countries—from huge economies such as the USA, China, India, the European Community and Brazil to small island countries.

Compared with the alternative of fossil fuels and other renewables, geothermal energy has substantial advantages in a social context. The most important are that it may provide a stable, domestic supply of energy, and that it is a 'clean' source of energy for both developing and developed countries. Although it is difficult to foresee exactly how these advantages weigh against the obstacles of uncertainty, attached by private investors as well as public entities and their financing institutions, studies strongly indicate the availability of options for more extensive development. In order to exploit the potential gains of geothermal energy by increasing applications of geothermal resources for electricity generation, countries must overcome these barriers and create or improve policies of sustainable renewable energies. They must integrate geothermal energy in their development plans, and decisions should not be made exclusively from a market point of view. The national authorities will have to be confident in their use of available instruments. Two domestic targets may be pointed out in this respect. One is to prepare a plan for an extensive coordination of the national electricity markets and to develop regional wholesale electricity markets. This limits the negative impacts of uncertainties both with respect to the markets, and to technological performance. The second is to internalize the social costs of the so-called negative externalities of energy production. One way to do this is to impose fines on activities that contribute to air pollution (this is being implemented in several countries as "carbon tax"). The social advantage of 'clean energy' thus becomes visible.

The true potential of geothermal sources (as renewable energy sources in general) will only be successfully harnessed to satisfy an increasing percentage of the fast growing energy demand, if governments prepare an institutional and regulatory framework that can overcome the present economic obstacles associated with geothermal projects, and promote geothermal energy and other 'clean' energy sources.

Despite uncertainties about the costs and benefits to the environment, the social gains from geothermal energy are substantial. The marginal benefits of 'clean' energy production may be as large as the total developing cost for a large, modern geothermal power plant. However, to take into account the environmental improvements in decision-making the national authorities, especially those of developing countries, will have to introduce incentives for private investors and for public electricity institutions and their financing entities to obtain the required international loans. Correspondingly, regulations, laws, and market instruments must be developed. 'Clean' energy projects may be promoted by tax incentives, subsidies, and investment guarantees. Also, investments in environmentally unsound projects must be made unattractive through special taxes, like charges on fossil fuels including user taxes and higher import taxes. Only such incentives will ensure that low fossil fuel prices do not result in public and private investments in conventional thermoelectric plants, and thereby become an obstacle for sustainable, geothermal energy development. Without incentives investors will select projects with the highest benefits from the purely market-economic value.

Worldwide privatization, or private sector participation in the energy sector is becoming more and more dominant. Private investment in the geothermal sector is very important due to the high costs and risks of exploration, and the high initial costs to develop a geothermal field. Due to the introduction of new and improved technologies, the costs associated with geothermal projects are declining and will continue to do so. Alternative energy sources vary widely in terms of cost, economies of scale, production properties, and the externalities of production. These properties are likely to be regarded differently depending on whether a private investor or a public authority makes the decision. In order to make private investors act in accordance with social interests it is now important, and may become even more important in the future, to find appropriate incentives for private investors.

A review of the economic properties of the geothermal energy production indicates that private investors may be reluctant about development, especially in developing countries. This can to some extent be confirmed from observations. High capital costs make the economy of geothermal energy plants more vulnerable to uncertainties in the energy market and to technology performance than alternative plants. Whereas the uncertainties in the energy markets affect all alternatives, the uncertainty about the technology or performance of the plant is probably larger for geothermal plants than for more traditional alternatives, such as thermoelectric energy production. It is well known from other sectors that this uncertainty significantly determines the choice of technology.

Additionally, one may also point out the clean development mechanism (CDM) of the Kyoto Protocol as one instrument that might encourage further development of geothermal energy. The CDM accounts for at least some of the positive environmental properties of geothermal energy. Moreover, it represents one possible way out of the constraints imposed by financial requirements. However, the uncertainties seem to represent an obstacle for active involvement to the investors in CDM projects. These uncertainties might be reduced if the technology becomes better known, but may also diminish if the management and coordination of CDM projects are lifted to an international level.

Finally, capacity building and popularization of geothermal energy will be required to create awareness and acceptance of it by politicians and decision makers. There is need for institutional strengthening, human resources formation, and consideration of geothermal projects as AIJ/CDM opportunities by lowering electricity generation costs (an important incentive for private investors).

The application of all these measures would improve the economic and sustainable development of the countries and regions by considering all ancillary benefits of geothermal energy, including energy security for the world largest economies and the social development of the poorest countries of our globe.

CHAPTER 4

Low-enthalpy resources as solution for power generation and global warming mitigation

> *"More recently, there has been a seismic shift in how climate change is perceived, and is widely considered to be the greatest market failure ever. This is in part due to the fact that many of the effects of climate change are beginning to manifest, and that the threats posed by continued warming will affect—and even possibly disrupt—the operation of markets, societies, ecosystems and cultures."*
>
> UNEP: Declaration on Climate Change by the
> Financial Services Sector, 2007.

4.1 OVERVIEW

Not only industrialized, but also developing countries, are compelled to implement the Kyoto protocol and the use of clean development mechanism (CDM). This is due to the constant increase in CO_2 content in the atmosphere and rise in the surface temperature of the globe. The global warming trend for the last 100 years is shown in Figure 4.1a, and the increase in the CO_2 level in the atmosphere is shown in Figure 4.1b. Under this situation, all developing countries are compelled to look for alternate CO_2-free sources of energy for all their developmental activities. Geothermal energy is the only option that can provide a safe, stable, and 'clean' energy source. Many alternative domestic renewable energy resources such as hydroelectric power and nuclear that comply with the regulations laid down by the United Nations Framework Convention for Climate Change (UNFCCC) have been increasingly questioned by society. Low-enthalpy geothermal resources must be targeted by developing countries to generate electric power as well as for direct applications. In all countries, urban populations are provided with their electricity requirements, but rural areas where greater than 70% of the world population live, need guaranteed electricity for socio-economic development. Fortunately, in many countries major low-enthalpy geothermal resources are located proximal to rural areas. Small-scale geothermal power plants are best suited to support socio-economic developmental activity in such regions. The rural areas in most of the developing countries are still using traditional energy sources like wood and cow dung that make the rural population vulnerable to a variety of respiratory diseases. Small-scale geothermal power plants can support electricity demand as well as create employment opportunities for the rural public.

The potential use of low-enthalpy geothermal energy for power production shall be analyzed here based on four different criteria: (1) its importance for countries which currently use predominantly imported fossil fuels for electricity generation; (2) its possibilities for countries with domestic available fossil fuel resources; (3) improvement of electricity supply in areas without access to grid electricity, and (4) its value for countries that use predominantly hydroelectric power.

The countries that meet the first three criteria produce most of their electricity from fossil fuels and would benefit from reduced CO_2 and other greenhouse gas emissions by replacing fossil fuels with environmentally friendly low-enthalpy geothermal energy resources thereby reducing the negative economic, social, and environmental impact to their countries while also helping to mitigate global warming. Benefits from emissions reduction and the CO_2 reduction potential by using low-temperature geothermal resources for power generation will be discussed in Chapter 4.2.

Countries that meet the first criteria depend on imported fossil fuels for their electricity generation. The use of domestic low-temperature geothermal resources would make them independent

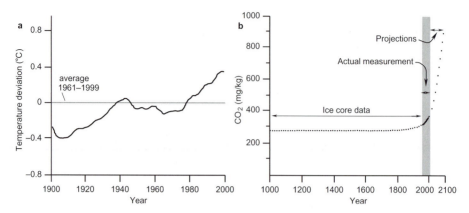

Figure 4.1.   (a) Global temperature deviation from the 1961–1999 average for the past 100 years and (b) past, present and future CO_2 concentration in the atmosphere (modified from UNFCCC 2006).

from fossil fuel imports eliminating the uncertainties of fluctuating fossil fuel prices. The lessened fossil fuel imports would contribute positively to the country's foreign trade balance, helping to reduce trade deficits by reducing external debts. This carries special importance for developing countries that have only limited financial resources, as it will be shown in more detail in Chapter 4.3.

The countries that meet criteria two, which have proper fossil fuel resources, can also benefit from using their low-temperature geothermal resources by saving their fossil fuel resources either for other applications or reserving them for the future. This is of importance for both developing and industrialized countries that are increasingly conscious about energy security.

The additional benefits for countries that meet criteria three are obvious: low-temperature geothermal resources can be tapped and used in areas that are remote from the national electricity grid-areas which do not have access to the electricity supply or which depend on expensive diesel generators. Depending on how far the fuel must be transported, prices as high as 0.50 US$/kWh are quite normal. This is principally important for rural electrification in not only developing countries (see section 4.5) but also in industrialized countries, particularly in remote areas.

Countries meeting the fourth criteria, which rely on hydroelectric power (e.g., Central/South America, Canada) can use geothermal energy to diversify their renewable energy thereby reducing their reliance on hydroelectric power, which is sensitive to climate events such as droughts, which affect the water supply causing blackouts (discussed in section 4.4).

4.2   BENEFITS THROUGH EMISSION REDUCTION

4.2.1   *The emission reduction potential*

Geothermal is a 'clean' energy source, which could significantly contribute to the reduction of greenhouse and other gas emissions by replacing fossil fuels for power generation. The emissions of CO_2, which is the principal greenhouse gas in the atmosphere, sulfur oxides (SO_x) and nitrogen oxides (NO_x) from a geothermal plant, are less than 2% of the emission of these gases by fossil fuel based power plants (UNFCCC 1997). On an average, a geothermal plant emits 0.893 kg CO_2/MWh (UNFCCC 1997), whereas coal, oil, and natural gas fired power plants emit 953, 817, and 193 kg CO_2/MWh, respectively (these are average values, which may vary locally due to variation of source composition) (Kasameyer 1997).

The economic, social and environmental value and importance of emissions reduction for the globe underline the benefits of geothermal energy. Several countries, especially those that make an effort to bring down emissions levels (to gain carbon credits and to avoid the carbon trade), have geothermal energy playing a significant role in their nation building process. Based on the forecast

(EIA 2007) for electricity production and source mix, the expected CO_2 emissions are calculated for different world regions and for certain specific countries, using the average emission values of Kasemeyer (1997). The results for the 2004–2030 period are given in Figures 4.2 and 4.3 and in Table 4.1, the source-specific and total CO_2 emissions are listed along with the average annual increases and the shares of CO_2 emissions from power plants of total anthropogenic CO_2 emissions. In accordance with the observations in Chapter 2, where we analyzed the future electricity generation changes and development of the source mix for the different world regions, we see clearly that CO_2 emissions are principally due to coal.

In 2004 coal fired power plants emitted worldwide 6407 million tonnes of CO_2 corresponding to 82% of CO_2 emitted by thermal power plants. CO_2 emissions from oil and gas were 766 and 624×10^6 tonnes (t), respectively accounting each for only about 10% of CO_2 released by fossil fuel power plants. Due to the increase in fossil fuel for electricity generation, by the year 2030 the CO_2 emissions from coal, oil, and gas fired power plants will increase by annual averages of 2.8, 0.9, and 3.3%, respectively amounting to 13,009, 979, and 1433×10^6 t, corresponding to 84, 6, and 9% of the total CO_2 emissions from power plants.

The CO_2 emission by power plants in OECD countries was 4247×10^6 t in 2004 (84% from coal) and is expected to increase at an average rate of 1.3% per year to reach 5865×10^6 t in the year 2030. Within the OECD regions the highest absolute and percentage increases of CO_2 emissions from power plants are found in the USA (increase of 1.9% per year from 2124×10^6 t to 3443×10^6 t), whereas in Europe the emission is expected to decrease (by –0.3% per year from 1186×10^6 t in 2004 to 1111×10^6 t on 2030). In the same period, in the non-OECD countries the CO_2 missions are expected to increase from 3547×10^6 t (80% from coal) in 2004 by an average of 3.9% per year to reach 9538×10^6 t (84% from coal) in 2030. Thereby the non-OECD Asia will be the largest contributor to CO_2 emissions from power plants. The emissions were 2436×10^6 t in 2004, much higher than those of the other non-OECD regions (Europe/Eurasia: 453×10^6 t, Middle East: 252×10^6 t, Africa: 281×10^6 t, and Central/South America: 127×10^6 t). In this region, China emitted about 1609×10^6 t in 2004 and contributed to 21% of the world CO_2 emission from power plants (USA 27%), but is expected to contribute to as much as 33% in 2030 (USA 22%, OECD Europe, 7%, Middle East 3%, Africa 4%, and Central/South America 1.5%).

So far we have estimated the CO_2 emission reduction potential, for a situation where fossil fuel is substituted by geothermal or other renewable energy resources. However, if we want to determine the potential of these measures as a contribution to global warming mitigation, we need to consider the CO_2 emissions from power plants (CO_{2_P}) in relation to the total anthropogenic CO_2 (CO_{2_T}) emissions of the respective regions or countries. In Table 4.1 we can see that CO_{2_P} contributed in 2004 with 29% to CO_{2_T} and the share is expected to increase to 36% in the year 2030. Lowest CO_{2_P}/CO_{2_T} ratios are found in the regions and countries, which use predominantly renewables for their power generation (Canada, Central/South America), but also Europe, where natural gas will be increasingly used for power generation, and in non-OECD Europe/Eurasia. Only in Europe, the CO_2 emission from power plants is expected to decrease significantly its share to total anthropogenic CO_2 emission until 2030. In the regions and countries with the highest absolute emission values, the CO_2 contributions from the sector to CO_{2_T} are expected to reach in 2030 about 43% in the USA, 44% in the non-OECD Asia region (due to the increasing share of coal, the most carbon-intensive of the fossil fuels), where high CO_2 emissions from power plants are expected in China and India with 46% and 55%, respectively (Table 4.1). As a consequence, by 2030, carbon dioxide emissions from the power plants from China and India combined are projected to account for 41% (only 27% in the year 2004) of the world total CO_2 emissions from power plants. From power plants in 2030, China alone will be responsible for 33% of the world CO_2 emission, whereas in 2004 it contributed only 21% of the world emission from power plants.

If we relate the regional CO_2 emissions from power plants to the global anthropogenic CO_2 emission (2004: $26,922 \times 10^6$ t, 2030: $42,880 \times 10^6$ t CO_2; EIA 2007), the power plants from the OECD regions contributed in 2004 to 16% of total anthropogenic CO_2 emissions and are projected to decrease slightly to reach 14% in 2030, whereas the contribution of non-OECD regions was only 13% in the year 2004, but will sharply increase to 22% in the year 2030 (Table 4.1). The projected

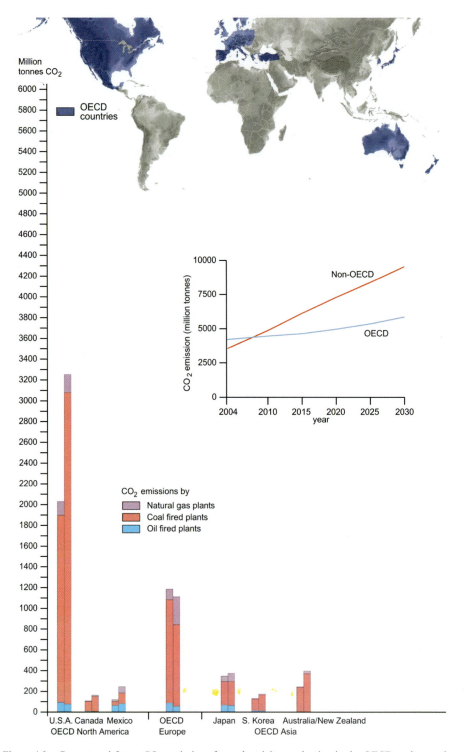

Figure 4.2.   Present and future CO_2 emissions from electricity production in the OECD regions estimated from power generation data of Figure 2.2 and average CO_2 emission values from Kasameyer 1997 for coal, oil, and gas fired power plants.

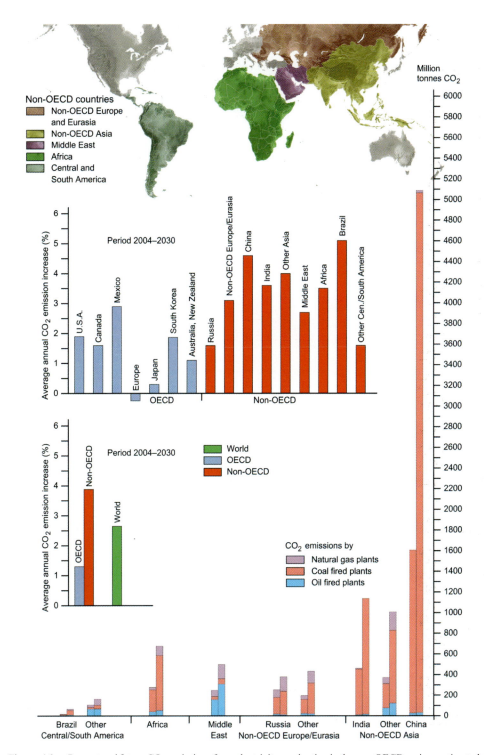

Figure 4.3.   Present and future CO_2 emissions from electricity production in the non-OECD regions estimated from power generation data of Figure 2.3 and average CO_2 emission values from (Kasameyer 1997) for coal, oil, and gas fired power plants.

Table 4.1. World CO_2 emissions from power generation (CO_{2P}) and comparison with total anthropogenic emission (CO_{2T}) values by region. The emission values from power generation are calculated from the different sources according to the energy mix and power generation data from EIA (2007) using average emission data from Kasameyer (1997) for coal, gas, and oil fired power plants; the total CO_2 emission data are from EIA (2007).

| | CO_2 emissions from power generation 10^6 tonnes CO_2 | | | | | | | | Increase (average) | Share of world CO_{2P} | | Share of CO_{2P} from CO_{2T} | | Share of CO_{2P} from world CO_{2T} | |
| | 2004 | | | | 2030 | | | | 2004–2030 | 2004 | 2030 | 2004 | 2030 | 2004 | 2030 |
	Coal	Oil	Gas	Total	Coal	Oil	Gas	Total	%/year	%	%	%	%	%	%
OECD	3579.5	339.9	327.5	4246.9	4916.5	315.4	633.4	5865.3	1.3	54.5	38.0	32.1	35.2	15.8	13.7
North America	2017.5	171.6	160.8	2349.8	3420.3	175.7	254.2	3850.2	1.9	30.1	25.0	34.1	41.0	8.7	9.0
USA	1886.0	99.7	138.0	2123.7	3173.5	87.4	181.8	3442.7	1.9	27.2	22.3	35.9	43.3	7.9	8.0
Canada	93.4	9.8	6.4	109.6	144.9	8.2	12.9	166.0	1.6	1.4	1.1	18.8	22.1	0.4	0.4
Mexico	38.1	61.3	16.4	115.8	102.0	80.1	59.3	241.3	2.9	1.5	1.6	30.1	34.5	0.4	0.6
Europe	1002.6	80.9	102.5	1185.9	787.2	54.7	269.0	1111.0	−0.3	15.2	7.2	27.1	23.7	4.4	2.6
Asia	559.4	87.4	64.3	711.1	708.1	85.0	110.2	903.3	0.9	9.1	5.9	32.6	35.2	2.6	2.1
Japan	228.7	67.8	52.3	348.8	233.5	63.7	79.3	376.5	0.3	4.5	2.4	27.6	28.8	1.3	0.9
S. Korea	144.9	18.8	7.7	171.4	243.0	19.6	15.1	277.7	1.9	2.2	1.8	34.5	40.2	0.6	0.6
Australia/New Zeal.	184.9	0.8	4.2	189.9	232.5	0.8	15.8	249.2	1.1	2.4	1.6	44.8	43.5	0.7	0.6
NON-OECD	2826.6	426.5	293.7	3546.8	8091.9	647.1	799.2	9538.2	3.9	45.5	61.9	26.3	36.4	13.2	22.2
Europe/Eurasia	301.1	40.0	111.7	452.9	528.0	33.5	253.8	815.3	2.3	5.8	5.3	16.1	21.0	1.7	1.9
Russia	163.9	17.2	74.1	255.2	229.7	12.3	139.5	381.5	1.6	3.3	2.5	15.1	17.5	0.9	0.9
Other	137.2	23.7	37.6	198.6	298.3	21.2	114.3	433.8	3.1	2.5	2.8	17.5	25.6	0.7	1.0
Asia	2253.8	112.7	69.5	2436.1	6889.2	169.1	243.0	7301.3	4.3	31.2	47.4	32.9	44.2	9.0	17.0
China	1580.1	27.0	2.1	1609.2	5067.1	30.2	24.3	5121.6	4.6	20.6	33.2	34.2	45.6	6.0	11.9
India	445.1	11.4	8.7	465.2	1118.8	16.3	40.0	1175.1	3.6	6.0	7.6	41.9	54.5	1.7	2.7
Other	229.7	74.3	58.5	362.5	702.4	121.7	178.7	1002.8	4.0	4.6	6.5	22.8	31.9	1.3	2.3
Middle East	30.5	158.5	62.9	251.9	51.5	311.3	139.7	502.5	2.7	3.2	3.3	19.5	21.8	0.9	1.2
Africa	216.3	40.9	24.1	281.3	539.4	53.1	91.7	684.2	3.5	3.6	4.4	30.6	41.3	1.0	1.6
Central/South America	24.8	74.3	27.8	126.9	82.9	80.1	71.2	234.2	2.4	1.6	1.5	12.4	12.7	0.5	0.5
Brazil	7.6	7.4	3.3	18.3	44.8	10.6	12.0	67.4	5.1	0.2	0.4	5.5	11.3	0.1	0.2
Other	17.2	67.0	24.3	108.5	38.1	70.3	59.3	167.6	1.7	1.4	1.1	15.7	13.4	0.4	0.4
TOTAL WORLD	6407.0	765.5	623.6	7796.1	13008.5	978.8	1432.6	15419.9	2.7	100.0	100.0	29.0	36.0	29.0	36.0

CO_2 contribution from the power plants to total global anthropogenic CO_2 emission by order of significance: non-OECD Asia 17% (thereby China 12 and India 3%), OECD North America 9% (USA 8%), OECD Europe 3%, and about 2% in each of the regions non-OECD Europe/Eurasia, OECD Asia, and Africa, whereas the lowest contributions come from Central/South America (0.5%).

These absolute numbers and the expected future trends towards an increasing contribution of CO_2 from power plants to total anthropogenic CO_2 emission are alarming and highlight the need to substitute high CO_2 emitting fossil fuels with low CO_2 emitting geothermal energy for power generation, whose resources, as discussed in Chapter 3, are large enough to produce the world entire power demand for both present and future generations.

Although over the past decade awareness of global climate change has been rapidly increasing, particularly in the wealthy regions of the world, the developing countries are clearly more concerned about their current problems. Most of them are reluctant to take an active part in the mitigation of climate change because they regard it as a problem created mainly by developed countries. On the other hand, it is recognized that global warming can not be mitigated unless developing countries also take an active part. Global warming is a global warning.

4.2.2 *The clean development mechanism (CDM) as incentive for developing countries*

In many countries climate change has received more attention by policy makers than any other environmental problem. So far, only developed countries are subject to proposed emission targets, but the Kyoto protocol, which entered into force on February 16, 2005, is open to participation from developing countries through the so-called 'clean development mechanism' (CDM). The idea is that developed countries can pay developing countries to reduce emissions, and thereby obtain a "credit" on their own emission targets (carbon credit). The motivation for the developing countries to become involved lies in the fact that their emissions of CO_2 and other greenhouse gases have suddenly been given a value, and may be traded. The emission cuts paid by the developed country will most likely comprise technology transfers to developing countries. Hence, the CDM provides foreign investors with additional motivations, and relaxes the financial restraints that seem to hamper the development of geothermal energy in many developing countries.

There are a lot of practical problems related to the initiation of CDM. It is, for example, difficult to determine the emission cuts of a particular project, because it requires a counterfactual assessment of future emissions. Moreover, the design and thereby the project cost are clearly subject to conflicts of interest between the investing country and the host country. Investing countries are concerned about emissions, while host countries are concerned about development. This also points to an inherent contradiction within the whole mechanism, because, if successful, it may contribute to growth, which encourages demand for energy, including that from fossil fuels. This generates emissions thereby defeating the very basic aim to reduce emissions. The parties of the convention are fully aware of this contradiction, recognizing that development in a large part of the world is necessary if climate change is to be mitigated in an effective way.

It is difficult to predict the importance of the CDM over the next 10–15 years, although it creates an incentive for producing energy without greenhouse gas emissions, and thereby contributes to enhancing the value of non-carbon energy. Equally promising is that some of the above mentioned obstacles for investment in domestic energy production in developing countries are reduced as a consequence of potential contributions from foreign interests. Financial constraints may become less limiting, since the investing country is to pay the full additional cost of the 'clean' alternative. Moreover, the uncertainties related to the technical performance of new technologies may be reduced, partly because investing countries in some cases may be familiar with it, and partly because the responsibility for operation and maintenance can be shared between the two countries. Thus, the development of 'clean' energy seems to fit particularly well with both aims of the CDM by reducing emissions of greenhouse gases and, as a consequence, reducing important obstacles for development in developing countries.

After the USA withdrew from the Kyoto protocol, the price of emission quotas in the international market is not expected to exceed 5 US\$/t CO_2 and under this condition can only be regarded as additional incentive. However, the CDM may at least reduce barriers to investing in capital-intensive geothermal plants. One of the uncertainties for investors is the uncertainty of the credit amounts to be achieved by investing in CDM projects. Ideally, CDM projects ought to replace existing sources of emissions as it is difficult to predict to what extent a new plant adds or replaces existing capacity. Hence, the investors may not get a full emissions credit, as this may partly depend on how important the conference of the parties of the UN Framework Convention on Climate Change will consider the development aspect of the CDM in comparison to aspects of greenhouse gas emission.

Despite its obvious advantages, geothermal energy has not attracted much attention as an opportunity for CDM projects so far. There are several relatively small AIJ (actions implemented jointly) and CDM projects on renewable energy related to hydroelectric power and wind energy. Only Nicaragua applied for a geothermal AIJ project (i.e., El Hoyo-Monte Galán geothermal field); it was approved but had to be cancelled (Bundschuh *et al*. 2002, 2007a).

The reasons for low investor interest may partly be related to elevated development costs of the geothermal field. As pointed out earlier, additional reasons could be overcome with an improved management of energy resources, not only from the national governments' side, but also in the context of managing CDM projects on the international stage. In its present premature phase, the CDM is predominantly a matter of bilateral cooperation between the investor and the host country. The international involvement applies mainly to the verification and control of projects. Thus, the full risk of each project is shouldered by the investor and the host country. This puts a particular burden on unconventional options, such as geothermal energy. This obstacle could, however, be reduced by pooling more and less uncertain CDM projects under the administration of a clearing house, with the aim of reducing the risk of single projects. Over the years the importance of international mechanisms similar to the CDM are likely to increase as well as become an incentive for the development of 'clean' energy resources. Despite current uncertainties about the willingness of many countries to take part in the Kyoto protocol and the following-up climate change mitigation agreements, one must expect that targets will be tightened and more countries will be subject to emission targets in the future. The price of quotas, and thereby the value of CDM projects, may increase as a result. An improved system for managing bilateral agreements may emerge thereby reducing the obstacles against implementing geothermal energy.

4.2.3   *Emission reduction benefits on a national level*

In addition to the contribution of global warming and climate change mitigation, national authorities have to consider the impacts of fossil fuel consumption within their national territories. The quality of the environment, air quality in particular, is closely related to the use of fossil fuels. A poor environmental standard does not necessarily affect the economic performance of industries directly, but may impose substantial costs to the society in terms of poor health, damage to vegetation (reducing agricultural yields), and buildings. The usual proposal to make private agents act in an environmentally acceptable manner is to impose restrictions on the sources of pollution, for example through imposing taxes. However, such restrictions are quite uncommon in developing countries. This is easy to understand because major polluters are often the large industries that represent the main potential players for future economic growth. To charge these industries for their self-caused pollution is considered a threat to growth. By turning to 'cleaner' energy, one may avoid the political impossibility of putting restrictions on the use of fossil fuels, and at the same time improve the quality of the environment. To assess the potential of such an option, all the efforts throughout the society need to be integrated.

Geothermal resources can provide a stable supply of energy, in contrast to many alternative domestic renewable energy resources, such as hydroelectric power. In comparison to thermoelectric plants, geothermal power plants require less maintenance and hence less interruption of production resulting in elevated output.

Although the importance of a stable energy supply is considered to be limited in developed countries with highly coordinated transmission systems, it may be vital to developing countries. We may consider India as an example with their fast emerging economy, resulting in a sharp increase of electricity demand, which is expected to nearly triple from the year 2004 to 2030. This expected demand is going to be fulfilled by burning an additional 20×10^9 t of coal. The Indian government has already permitted the lease for opening new coal mines and increasing the production of coal from the existing mines (Chandrasekharam *et al.* 2006, Chandrasekharam and Chandrasekhar 2007). This country predominantly uses coal to produce its power: 467×10^6 MWh in the year 2004, and forecasted 1174×10^6 MWh in the year 2030 (annual average increase 3.6%), corresponding to 74 and 67% of the national power generation, respectively (EIA 2007). This 3-fold increase of coal combustion for power generation results in a respective huge increase of CO_2 emissions from 445×10^6 t in 2004 to estimated 1119×10^6 t in the year 2030 (EIA 2007 and Chandrasekharam and Chandrasekhar 2006) and other greenhouse gases. These gases contribute not only significantly to global warming with all its negative effects on the climate, reduction of crop production, etc., but also severely affect the national territory through air contamination, which is a burden on the public health sector, and reduce the labor power of the country. To avoid or to reduce these negative effects, alternative environmentally friendly power sources need to be considered. Geothermal resources offer such alternatives. However, the actual geothermal potential of this country has not yet been estimated and the figures that are reported in the literature are based only on the surface emissions and not based on the aquifer parameters of individual geothermal provinces (Chandrasekharam 2005). As a working number it can be assumed that the potential of low- and high-enthalpy geothermal resources is greater than by a factor of two to three considering the presently reported amount of 10,000 MW_e (Chandrasekharam 2005). While this source is not yet utilized, plans are underway to tap this huge resource in the coming decade.

But why does India not develop its huge geothermal resources as a 'clean' and stable source for meeting its power demands? The answer lies in the 192×10^9 t recoverable coal reserves, which is encouraging coal based power projects and hampering the healthy growth of non-conventional energy programs. As mentioned above, the country is keen in providing additional licenses to open new coal mines and increase the coal production rather than considering geothermal resources in the national energy planning. Furthermore, neither the government bodies nor the independent power producers (IPPs) are aware of the vast available geothermal energy resource in the country. As mentioned above, additional use of coal to meet the future energy demand is detrimental to India as the estimated CO_2 emission is going to increase in the 2005–2030 period by a factor of about 3 (see Table 4.1 and Chandrasekharam and Chandrasekhar 2006). In India the present day cost of one unit (kWh) of power is less than ~0.02 US$ in the case of coal based power, liquid fuel based power costs ~0.04 US$, and hydroelectric power costs ~0.03 US$ (Chandrasekharam 2001a). But the expenditure spent to meet the consequences (like disposal of fly ash; treating the coal with high ash content, providing 'clean' air to the population, and meeting pollution related health hazards, etc.) is high, which automatically increases 0.02 US$ a unit to several cents. Now a time has come to reexamine those alternate energy sources such as geothermal, which were not viable a decade ago due to non-availabilities of advanced technical know how.

4.3 BENEFITS OF DOMESTIC GEOTHERMAL RESOURCES *VERSUS* FOSSIL FUEL IMPORTS

There are two types of countries which are importing fossil fuels. The first group imports fossil fuels because it has no domestic fossil fuel resources, or they are not economically accessible. The second group is comprised of mostly industrialized countries which have significant domestic fossil fuel resources, but attempt not to use them in order to keep them as reserves for future use. For both of these country groups, the vast potential of domestic low-temperature geothermal resources, which are found practically everywhere, and the locally limited high-temperature resources, are excellent solutions to avoid or to reduce fossil fuel imports and all the related problems, which will be discussed in section 4.3.2 in more detail, and to improve or guarantee energy security of their countries.

Table 4.2 gives a country by country overview of the total generated power, the source mix, and the percentage of the population which has access to electricity. This table includes all countries with the exception of those islands whose electricity is generated by >90% from fossil fuels, and those countries whose power produced is >80% from hydroelectric power. The countries from both these groups are listed separately in the Tables 4.3 and 4.4, respectively.

4.3.1   *Benefits of geothermal for countries without fossil fuel resources*

In Chapter 2, it is stated that in the developing countries, which actually depend on the availability and the costs and uncertainties of fossil fuels, the electricity demand will be growing exponentially during the next few decades and it is predicted that fossil fuels will have the highest growth. It is also predicted that, especially in countries without fossil fuel resources, the demand-supply ratio will greatly depend on the availability, cost, and uncertainties related to fossil fuel products, calling for the development of alternative domestic reliable and more sustainable energy resources such as the low-temperature geothermal resources discussed above.

The development, especially of the low-temperature geothermal resources, could help reduce growing trade deficit and external debts. Both will expand as the demand for electricity continues to increase, unless the energy mix for electricity production is changed to alternative domestic energy resources.

Some developed countries have gone ahead in utilizing these sources for power generation, however, no developing country is using this energy resource for commercial electricity generation (with exception of China, which has a small power plant using low-temperature geothermal resources; see Chapter 10). Therefore, it is necessary to encourage the development of renewable domestic geothermal energy resources in order to mitigate vulnerability to world energy markets.

The important economic benefits of geothermal energy development for many countries of the world can be seen by considering Nicaragua, as an example. Nicaragua is one of the lowest-income countries in Latin America. Nicaragua spends about US\$ 95,000,000 per year to import oil for electricity generation, which amounts to about 4% of its GDP (73% of the installed capacity corresponds to thermal plants) (Bundschuh *et al.* 2007a). Geothermal power plants could promptly displace all thermal plants and improve the country's foreign trade balance. The development of Nicaragua's geothermal resources would not only solve its energy demand, but would also allow the country to supply electricity to its neighbouring countries.

Another case from Central America is that of Guatemala, which could replace its thermal power plants used for power supply with geothermal units and export the locally produced oil, to improve its foreign trade balance, or use it to refine transportation fuel.

Also for many islands, especially small ones, low- as well as high-enthalpy geothermal resources can become an ideal domestic energy source. As seen from Table 4.3, it is apparent that most of them depend exclusively or predominantly on imported fossil fuel resources for power generation. Only some islands with favorable geomorphological and climatic conditions are using hydroelectric power. These are, in general, larger islands whose fossil fuel shares for power generation are given in the following (the remaining percent corresponds to renewables, predominantly to hydropower): In the Caribbean: Haiti (60%), Dominica (47%), and Saint Vincent and the Grenadines (69%), in the Atlantic the Faroe islands (62%), and Sao Tome and Principe (41%), in the Indian Ocean the island of Madagascar (36%), and in the Pacific/Oceania region, the Fiji islands (19%), French Polynesia (61%), New Caledonia (76%), and Samoa (58%) (Table 4.2). Iceland, which uses only 0.1% fossil fuels, 83% geothermal and 18% hydroelectric power is an exception. The other islands, which use over 90% imported fossil fuels are listed in Table 4.3, many of them depend on 100% of fossil fuel imports, predominantly of oil. For many of these islands, domestic geothermal resources are an ideal source for power generation since many of them are located within geothermally active zones, so that the probability of obtaining low-enthalpy fluids (minimum ~80 °C) at shallow depths is high.

4.3.2   *Problems related to fossil fuel imports*

Increasing energy demand may always be covered by importing fossil fuels, but this makes countries vulnerable to shifts in world energy markets, which have oscillated vigorously over the past 5 decades. Developing countries seldom possess stable enough economic foundations to

Table 4.2. Countries (in alphabetical order), with population, access to electricity, national electricity generation as total and by source; for comparison 2001 and 2005 geothermal energy shares are also given (Bertani 2005). This table includes all countries except islands with >90% power generation from fossil fuels, and countries with >80% from hydroelectric power, which are listed separately in Tables 4.3 and 4.4, respectively.

Country (or dependency)	Population 2007 est.[1]	Electr. access 2000[2] %	Electr. generat.[1] 10^9 kWh 2004	Electricity source[1] year 2001%					Geot 2005
				fos	nuc	hyd	geot	other	
Afghanistan	31889923	2	0.734	36.3	0.0	63.7	0.0	0.0	0.0
Algeria	33333216	98	29.390[3]	99.7	0.0	0.3	0.0	0.0	0.0
Angola	12263596	12	2.194	36.4	0.0	63.6	0.0	0.0	0.0
Argentina	40301927	94.6	93.940	52.2	6.7	40.8	0.0	0.2	0.0
Armenia	2971650	na	6.317[6]	42.3	30.7	27.0	0.0	0.0	0.0
Australia	20434176	100	225.300	90.8	0.0	8.3	0.0	0.9	0.0
Austria	8199783	100	64.900[4]	29.3	0.0	67.2	0.0	3.5	0.0
Azerbaijan	8120247	na	20.350	89.7	0.0	10.3	0.0	0.0	0.0
Bahrain	708573	99.4	7.794	100.0	0.0	0.0	0.0	0.0	0.0
Bangladesh	150448339	20.4	18.090	93.7	0.0	6.3	0.0	0.0	0.0
Belarus	9724723	na	29.330	99.5	0.0	0.1	0.0	0.4	0.0
Belgium	10392226	100	80.220	38.4	59.3	0.6	0.0	1.8	0.0
Belize	294385	na	0.175	59.9	0.0	40.1	0.0	0.0	0.0
Bolivia	9119152	60.4	4.472	44.4	0.0	54.0	0.0	1.5	0.0
Bosnia & Herzegovina	4552189	na	12.980	53.5	0.0	46.5	0.0	0.0	0.0
Brunei	374577	na	2.913[6]	100.0	0.0	0.0	0.0	0.0	0.0
Bulgaria	7322858	na	45.700[7]	47.8	44.1	8.1	0.0	0.0	0.0
Burkina Faso	14326203	13	0.400	69.9	0.0	30.1	0.0	0.0	0.0
Burma	47373958	na	6.020[7]	44.5	0.0	43.4	0.0	12.1	0.0
Cambodia	13995904	15.8	0.131	65.0	0.0	35.0	0.0	0.0	0.0
Canada	33390141	100	573.000	28.0	12.8	57.9	0.0	1.3	0.0
Chad	9885661	na	0.094	100.0	0.0	0.0	0.0	0.0	0.0
Chile	16284741	99	47.600[7]	47.0	0.0	51.5	0.0	1.4	0.0
China	1321851888	98.6	2494.000	80.2	1.2	18.5	0.0	0.1	0.0
Colombia	44379598	81	46.930	26.0	0.0	72.7	0.0	1.3	0.0
Ivory Coast	18013409	50	4.625	61.9	0.0	38.1	0.0	0.0	0.0
Croatia	4493312	na	12.950	33.6	0.0	66.0	0.0	0.4	0.0
Czech Rep.	10228744	100	79.140	76.1	20.0	2.9	0.0	1.0	0.0
Denmark	5468120	na	43.350[7]	82.7	0.0	0.1	0.0	17.3	0.0
Djibouti	496374	na	0.200	100.0	0.0	0.0	0.0	0.0	0.0
Dominica	72386	na	0.084	47.1	0.0	52.9	0.0	0.0	0.0
Ecuador	13755680	80	12.200	81.0	0.0	19.0	0.0	0.0	0.0
Egypt	80335036	93.8	91.720	81.0	0.0	19.0	0.0	0.0	0.0
El Salvador	6948073	70.8	5.293[7]	44.0	0.0	30.9	22.2	2.9	22.0
Eq. Guinea	551201	na	0.026	94.3	0.0	5.7	0.0	0.0	0.0
Eritrea	4906585	17	0.276	100.0	0.0	0.0	0.0	0.0	0.0
Estonia	1315912	na	9.290	99.8	0.0	0.1	0.0	0.2	0.0
Faroe islands	45511	na	0.293	62.4	0.0	37.6	0.0	0.0	0.0
Finland	5238460	100	81.600	39.0	30.4	18.7	0.0	11.8	0.0
France	63713926	100	549.400[5]	8.2	77.1	14.0	0.0	0.0	0.0
French Polynesia	278963	na	0.607	60.7	0.0	39.3	0.0	0.0	0.0
Gabon	1454867	31	1.543	34.5	0.0	65.5	0.0	0.0	0.0
Gambia	1688359	na	0.145	100.0	0.0	0.0	0.0	0.0	0.0
Germany	82400996	100	566.900	61.8	29.9	4.2	0.0	4.1	0.0
Greece	10706290	100	55.510	94.5	0.0	3.8	0.0	1.7	0.0
Guatemala	12728111	66.7	7.200[6]	51.9	0.0	35.2	3.7	9.2	3.0

(continued)

Table 4.2   *(continued)*

Country (or dependency)	Population 2007 est.[1]	Electr. access 2000[2] %	Electr. generat.[1] 10^9 kWh 2004	fos	nuc	hyd	geot	other	Geot 2005
Guinea	9947814	na	0.840[7]	45.5	0.0	54.5	0.0	0.0	0.0
Guinea-Bissau	1472780	na	0.058	100.0	0.0	0.0	0.0	0.0	0.0
Guyana	769095	na	0.819	99.4	0.0	0.6	0.0	0.0	0.0
Haiti	8706497	34	0.536	60.3	0.0	39.7	0.0	0.0	0.0
Honduras	7483763	54.5	4.805[6]	50.2	0.0	49.8	0.0	0.0	0.0
Hong Kong	6980412	na	38.450[6]	100.0	0.0	0.0	0.0	0.0	0.0
Hungary	9956108	100	31.830	60.1	39.0	0.5	0.0	0.3	0.0
India	1129866154	43	630.600	81.7	3.4	14.5	0.0	0.3	0.0
Indonesia	234693997	53.4	123.400[4]	84.4	0.0	10.5	5.1	0.0	6.7
Iran	65397521	97.9	155.700	97.1	0.0	2.9	0.0	0.0	0.0
Iraq	27499638	95	34.600[7]	98.4	0.0	1.6	0.0	0.0	0.0
Ireland	4109086	100	23.260	95.9	0.0	2.3	0.0	1.7	0.0
Israel	6426679	100	46.070	99.9	0.0	0.1	0.0	0.0	0.0
Italy	58147733	100	277.600	78.6	0.0	18.4	1.7	1.3	1.9
Japan	127433494	100	996.000[6]	60.0	29.8	8.4	0.4	1.4	0.3
Jordan	6053193	95	8.431	99.4	0.0	0.6	0.0	0.0	0.0
Kazakhstan	15284929	na	66.500[5]	84.3	0.0	15.7	0.0	0.0	0.0
Kenya	36913721	7.9	5.709	17.7	0.0	71.0	8.7	2.6	19.2
Korea, North	23301725	20	21.710	29.0	0.0	71.0	0.0	0.0	0.0
Korea, South	49044790	100	345.200	62.4	36.6	0.8	0.0	0.0	0.0
Kuwait	2505559	na	40.370	100.0	0.0	0.0	0.0	0.0	0.0
Latvia	2259810	na	4.550	29.1	0.0	70.9	0.0	0.0	0.0
Lebanon	3925502	95	9.762	97.2	0.0	2.8	0.0	0.0	0.0
Liberia	3195931	na	0.325	100.0	0.0	0.0	0.0	0.0	0.0
Libya	6036914	99.8	19.440	100.0	0.0	0.0	0.0	0.0	0.0
Lithuania	3675439	na	17.800	16.5	77.7	5.7	0.0	0.0	0.0
Luxembourg	480222	100	3.203[4]	57.3	0.0	25.2	0.0	17.5	0.0
Macau	456989	na	2.027[6]	100.0	0.0	0.0	0.0	0.0	0.0
Macedonia	2055915	na	5.935[7]	83.7	0.0	16.3	0.0	0.0	0.0
Madagascar	19448815	8	0.984	36.1	0.0	63.9	0.0	0.0	0.0
Malaysia	24821286	96.9	78.240	89.5	0.0	10.5	0.0	0.0	0.0
Mali	11995402	na	0.410	41.7	0.0	58.3	0.0	0.0	0.0
Mauritania	3270065	na	0.177	85.9	0.0	14.1	0.0	0.0	0.0
Mexico	108700891	95	242.400	78.4	4.2	14.2	3.2	0.0	3.1
Moldova	4320490	na	1.229[6]	90.6	0.0	9.4	0.0	0.0	0.0
Mongolia	2951786	90	3.430[7]	100.0	0.0	0.0	0.0	0.0	0.0
Morocco	33757175	71.1	18.480	95.4	0.0	4.6	0.0	0.0	0.0
Namibia	2055080	34	1.397	na	na	na	0.0	na	0.0
Netherlands	16570613	100	92.700	89.9	4.3	0.1	0.0	5.7	0.0
N. Caledonia	221943	na	1.675	76.3	0.0	23.7	0.0	0.0	0.0
New Zealand	4115771	100	41.100	31.6	0.0	57.8	6.1	10.7	7.1
Nicaragua	5675356	48	2.778[7]	83.9	0.0	7.7	6.1	2.3	9.8
Niger	12894865	na	0.232	100.0	0.0	0.0	0.0	0.0	0.0
Nigeria	135031164	40	19.060	61.9	0.0	38.1	0.0	0.0	0.0
Oman	3204897	94	14.330	100.0	0.0	0.0	0.0	0.0	0.0
Pakistan	164741929	52.9	80.240	68.8	3.0	28.2	0.0	0.0	0.0
Panama	3242173	76.1	6.888	37.0	0.0	61.3	0.0	1.7	0.0

(continued)

Table 4.2   *(continued)*

Country (or dependency)	Population 2007 est.[1]	Electr. access 2000[2] %	Electr. generat.[1] 10^9 kWh 2004	Electricity source[1] year 2001%					Geot 2005
				fos	nuc	hyd	geot	other	
Papua New Guinea	5795887	na	3.358	54.1	0.0	45.9	0.0	0.0	na
Philippines	91077287	87.4	56.570[6]	55.6	0.0	17.5	21.5	5.4	19.1
Poland	38518241	100	143.500	98.1	0.0	1.5	0.0	0.4	0.0
Portugal	10642836	100	42.520	64.5	0.0	41.3	0.0	4.1	0.0
Qatar	907229	95	12.400	100.0	0.0	0.0	0.0	0.0	0.0
Romania	22276056	na	54.530	62.5	9.9	27.6	0.0	0.0	0.0
Russia	147377752	na	952.400[6]	66.3	16.4	17.2	0.0	0.1	0.0
St. Vincent & Grenadines	118149	na	0.114	69.3	0.0	30.7	0.0	0.0	0.0
Samoa	214265	na	0.108	58.0	0.0	42.0	0.0	0.0	0.0
Sao Tome & Principe	199579	na	0.018	41.2	0.0	58.8	0.0	0.0	0.0
Saudi Arabia	26601038	97.7	155.200	100.0	0.0	0.0	0.0	0.0	0.0
Senegal	12521851	na	1.453	100.0	0.0	0.0	0.0	0.0	0.0
Serbia	10150265	na	33.870	na	na	na	0.0	na	0.0
Sierra Leone	6144562	na	0.244	100.0	0.0	0.0	0.0	0.0	0.0
Singapore	4533009	100	32.640	100.0	0.0	0.0	0.0	0.0	0.0
Slovakia	5447502	100	31.290[6]	30.3	53.6	16.0	0.0	0.0	0.0
Slovenia	2009245	na	14.900[7]	35.2	36.8	27.3	0.0	0.7	0.0
Somalia	9118773	na	0.269	100.0	0.0	0.0	0.0	0.0	0.0
South Africa	43997828	66.1	227.200	93.5	5.5	1.1	0.0	0.0	0.0
Spain	40448191	100	263.300	50.4	27.2	18.2	0.0	4.1	0.0
Sri Lanka	20926315	62	8.766[6]	51.7	0.0	48.3	0.0	0.0	0.0
Sudan	39379358	30	3.576	52.1	0.0	47.9	0.0	0.0	0.0
Suriname	470784	na	1.509	25.2	0.0	74.8	0.0	0.0	0.0
Swaziland	1133066	na	0.156[6]	58.0	0.0	42.0	0.0	0.0	0.0
Sweden	9031088	100	150.500	4.0	43.0	50.8	0.0	2.3	0.0
Switzerland	7554661	100	61.970	1.3	37.1	59.5	0.0	2.0	0.0
Syria	19314747	85.9	34.940[4]	57.6	0.0	42.4	0.0	0.0	0.0
Taiwan	22858872	na	189.700[6]	71.4	22.6	6.0	0.0	0.0	0.0
Thailand	65068149	82.1	121.700	91.3	0.0	6.4	0.0	2.4	0.0
Togo	5701579	9	0.286	98.7	0.0	1.3	0.0	0.0	0.0
Tunisia	10276158	94.6	11.810	99.5	0.0	0.5	0.0	0.0	0.0
Turkey	71158647	95	143.300	79.3	0.0	20.4	0.0	0.3	0.0
Turkmenistan	5097028	na	10.790[3]	99.9	0.0	0.1	0.0	0.0	0.0
Ukraine	46299862	na	192.100[7]	48.6	43.5	7.9	0.0	0.0	0.0
United Arab Emirates	4444011	96	49.520	100.0	0.0	0.0	0.0	0.0	0.0
United Kingdom	60776238	100	363.200	73.8	23.7	0.9	0.0	1.6	0.0
USA	301139947	100	3979.000	71.4	20.7	5.6	0.4	1.9	0.5
Uzbekistan	27780059	na	49.000	88.2	0.0	11.8	0.0	0.0	0.0
Venezuela	26023528	94	93.030	31.7	0.0	68.3	0.0	0.0	0.0
Vietnam	82265356	75.8	40.110	43.7	0.0	56.3	0.0	0.0	0.0
West Bank	2535927	na	na	100.0	0.0	0.0	0.0	0.0	0.0
West. Sahara	382617	na	0.085	100.0	0.0	0.0	0.0	0.0	0.0
Yemen	22230531	50	4.077[3]	100.0	0.0	0.0	0.0	0.0	0.0
Zimbabwe	12311143	39.7	9.412	47.0	0.0	53.0	0.0	0.0	0.0

[1] CIA 2007; [2] WRI 2007; [3] 2004 estimate; [4] 2005 estimate; [5] 2006 estimate; [6] 2005; [7] 2006; na: not available.

Table 4.3   Islands, which predominantly (>90%) use imported fossil fuels, mostly oil for electricity generation: Population numbers, percentage of people who have access to electricity, national electricity generation as total and by source.

Country (or dependency)	Popul. 2007 est.[1]	Electr. access[2] 2000 %	Electr. generat.[1] 10^9 kWh 2004	Electricity source[1] year 2001%				
				fos	nuc	hyd	geot	other
American Samoa	57663	na	0.128	100.0	0.0	0.0	0.0	0.0
Antigua & Barbuda	69481	na	0.105	100.0	0.0	0.0	0.0	0.0
Bahamas, The	305655	na	1.795	100.0	0.0	0.0	0.0	0.0
Barbados	280946	na	0.833	100.0	0.0	0.0	0.0	0.0
Bermuda	66163	na	0.683	100.0	0.0	0.0	0.0	0.0
British Virgin islands	23552	na	0.042	100.0	0.0	0.0	0.0	0.0
Cape Verde	423613	na	0.044	100.0	0.0	0.0	0.0	0.0
Cayman islands	46600	na	0.400	100.0	0.0	0.0	0.0	0.0
Cook islands	21750	na	0.028	100.0	0.0	0.0	0.0	0.0
Cyprus	788457	na	4.448	100.0	0.0	0.0	0.0	0.0
East Timor	1084971	na	na	100.0	0.0	0.0	0.0	0.0
Falkland islands	3105	na	0.016	100.0	0.0	0.0	0.0	0.0
Greenland	56344	na	0.295	100.0	0.0	0.0	0.0	0.0
Grenada	87971	na	0.171	100.0	0.0	0.0	0.0	0.0
Guam	173456	na	1.764	100.0	0.0	0.0	0.0	0.0
Kiribati	107817	na	0.013	100.0	0.0	0.0	0.0	0.0
Maldives	369031	na	0.150	100.0	0.0	0.0	0.0	0.0
Malta	401880	na	2.291	100.0	0.0	0.0	0.0	0.0
Montserrat	9538	na	0.002[3]	100.0	0.0	0.0	0.0	0.0
Nauru	13528	na	0.030	100.0	0.0	0.0	0.0	0.0
Netherlands Antilles	223652	na	1.005	100.0	0.0	0.0	0.0	0.0
Niue	1492	na	0.003	100.0	0.0	0.0	0.0	0.0
Pitcairn islands	48	na	na	100.0	0.0	0.0	0.0	0.0
Saint Helena	7543	na	0.007	100.0	0.0	0.0	0.0	0.0
Saint Kitts & Nevis	39349	na	0.125	100.0	0.0	0.0	0.0	0.0
Saint Lucia	170649	na	0.290	100.0	0.0	0.0	0.0	0.0
Saint Pierre & Miquelon	7036	na	0.050	100.0	0.0	0.0	0.0	0.0
Seychelles	81895	na	0.208	100.0	0.0	0.0	0.0	0.0
Solomon islands	566842	na	0.055	100.0	0.0	0.0	0.0	0.0
Tonga	116921	na	0.041	100.0	0.0	0.0	0.0	0.0
Turks and Caicos is.	21746	na	0.007	100.0	0.0	0.0	0.0	0.0
Vanuatu	211971	na	0.043	100.0	0.0	0.0	0.0	0.0
Virgin islands	108448	na	0.980	100.0	0.0	0.0	0.0	0.0
Trinidad & Tobago	1056608	99	6.049	99.8	0.0	0.0	0.0	0.2
Puerto Rico	3944259	na	24.140	99.2	0.0	0.8	0.0	0.0
Jamaica	2780132	90	6.913	96.8	0.0	1.8	0.0	1.4
Cuba[5]	11394043	97	14.100[4]	93.9	0.0	0.6	0.0	5.4
Dominican Republic	9365818	67	15.020	92.0	0.0	7.6	0.0	0.4
Mauritius	1250882	100	2.107	90.8	0.0	9.2	0.0	0.0
Comoros	711417	na	0.019	90.6	0.0	9.4	0.0	0.0

[1]CIA 2007; [2]WRI 2007; [3]2003; [4]2005; [5]Cuba has some domestic gas and oil resources, however most fossil fuels are imported; na: not available.

afford energy import increases. The challenge is to find sources that can be provided domestically without large environmental impacts. Within the process of development, alternative energy sources cannot be evaluated only with reference to expected monetary costs and benefits. One must also take into account that a stable supply of energy is a key factor in the transition from a high-risk, low-income country to a developed, better integrated country in the world economy.

Table 4.4   Countries, which cover over 80% of their electricity by hydropower: Population numbers, percentage of people who have access to electricity, national electricity generation as total and by source; for comparison additionally to the 2001 geothermal shares also those of the year 2005 are given (Bertani 2005).

| Country (or dependency) | Population 2007 est.[1] | Electr. access[2] 2000 % | Electr. generat.[1] 10^9 kWh 2004 | Electricity source[1] year 2001% | | | | | Geot 2005 |
				fos	nuc	hyd	geot	other	
Bhutan	2327849	na	2.050	0.1	0.0	99.9	0.0	0.0	0.0
Paraguay	6669086	75.7	51.770	0.0	0.0	99.9	0.0	0.1	0.0
Congo, Rep.	3800610	21	6.847[3]	0.3	0.0	99.7	0.0	0.0	0.0
Zambia	11477447	12	9.962	0.5	0.0	99.5	0.0	0.0	0.0
Burundi	8390505	na	0.137	0.6	0.0	99.4	0.0	0.0	0.0
Norway	4627926	100	108.900	0.4	0.0	99.3	0.0	0.0	0.0
Uganda	30262610	3.7	1.894	0.9	0.0	99.1	0.0	0.0	0.0
Uruguay	3460607	98	8.183	0.7	0.0	99.1	0.0	0.3	0.0
Laos	6521998	na	3.936	1.4	0.0	98.6	0.0	0.0	0.0
Congo, Dem. Rep.	65751512	6.7	0.353	1.8	0.0	98.2	0.0	0.0	0.0
Tajikistan	7076598	na	16.500	1.9	0.0	98.1	0.0	0.0	0.0
Rwanda	9907509	na	0.093	2.3	0.0	97.7	0.0	0.0	0.0
Cameroon	18060382	20	3.924	2.7	0.0	97.3	0.0	0.0	0.0
Albania	3600523	na	5.451	2.9	0.0	97.1	0.0	0.0	0.0
Mozambique	20905585	7.2	11.580	2.9	0.0	97.1	0.0	0.0	0.0
Ethiopia	76511887	4.7	2.294[3]	1.3	0.0	96.9	1.9	0.0	0.0
Malawi	13603181	5	1.293	3.3	0.0	96.7	0.0	0.0	0.0
Ghana	22931299	45	6.489	5.0	0.0	95.0	0.0	0.0	0.0
Kyrgyzstan	5284149	na	14.060	7.6	0.0	92.4	0.0	0.0	0.0
Nepal	28901790	15.4	2.511	8.5	0.0	91.5	0.0	0.0	0.0
Benin	8078314	22	0.082	14.2	0.0	85.8	0.0	0.0	0.0
Peru	28674757	73	23.990[4]	14.5	0.0	84.7	0.0	0.8	0.0
Brazil	190010647	94.9	546.000	8.3	4.4	82.7	0.0	4.6	0.0
Iceland	301931	100	8.474	0.1	0.0	82.5	14.7	2.8	17.2
Costa Rica	4133884	95.7	8.400	1.5	0.0	81.9	14.1	2.5	15.0
Fiji	918675	na	0.817	18.5	0.0	81.5	0.0	0.0	0.0
Tanzania	39384223	10.5	2.562	18.9	0.0	81.1	0.0	0.0	0.0
Georgia	4646003	na	6.804	19.7	0.0	80.3	0.0	0.0	0.0
Central African R.	4369038	na	0.109	19.8	0.0	80.2	0.0	0.0	0.0

[1] CIA 2007; [2] WRI 2007; [3] 2006 estimate; [4] 2004 estimate; na: not stated.

Although attention is drawn to integrating a wider scope of social impacts when evaluating alternative energy sources, the importance of monetary costs and benefits of investments in developing countries should not be underemphasized. Developing countries are often faced with financial constraints, unlike common to developed countries. It may not be sufficient to point out social benefits of 'clean' energy sources, even when tangible, such as in the case of reduced dependency of energy imports, if these benefits do not enter into the accounting of the fundraiser.

Among the social benefits of alternative non-conventional energy resources, enhanced security of the energy supply, and restricted environmental impacts, may be pointed out as particularly important. In a purely economic sense, the risk of increased dependency on energy imports may be translated to uncertainty in the price of fossil fuels. In a national context, the consequences of a sudden energy supply shortage that would lead to a rapid increase in prices may have serious indirect impacts on the total economy, especially if subject to financial constraints.

This became particularly evident during the rise of oil prices in the 1970s, when most countries of the world faced an acute trade deficit. The lowest-income countries suffered the most because their economies were not able to use alternative energy, or adjust their activities to a less energy intensive mode. At the same time, the loss of export income as a result of the general slow-down of global economic growth, accelerated the initial problems of more expensive energy. The social dynamics of these problems made it very difficult for developing countries to recover. Ten years later, the drop in energy prices did not have the reverse effect, neither on the world economy, nor as a catalyst for growth in developing countries.

The difference between the perspectives of private investors and public authorities is striking. To a private investor, the uncertainties in the fossil fuel markets mean uncertain cost of imports, but, at the same time, a corresponding uncertainty in the value of the non-conventional alternative. If fossil fuels are expensive, the value of the domestic alternative is high. But if the world market price of fossil fuel is low, the domestic alternative has also a low value. Therefore, fluctuations in world markets for fossil fuels do not necessarily affect the relative values between fossil fuels and other energy sources because they are all affected by the same uncertainty. But other factors, which are important to investors, may turn out to be disadvantageous to the domestic alternatives. These factors include high capital costs of alternative energy production, which involves a considerable risk to the investor, and uncertainties about the performance of the technology.

On the other hand, the importing of fossil fuels involves disadvantages to the national authorities, which private investors may disregard. In a national context, high prices of fossil fuels in the world markets are in most cases considered disadvantageous to the national economy. High prices may lead to long-term recession, which is reinforced if the economy is dependent on energy imports. Therefore, domestic energy resources, such as geothermal, may contribute to stabilizing the domestic energy market. At the same time, it is harder to believe that low prices in the world market spur development impulses, because all countries are subject to low prices. Countries depending on energy imports do not gain from low energy prices relative to other countries. The interpretation of "good news" and "bad news" is likely to be opposite for a private investor and an agent that makes decisions on behalf of a nation. This explains why securing an energy supply may be considered a separate issue at the national level.

4.4   BENEFITS OF GEOTHERMAL *VERSUS* HYDROELECTRIC POWER GENERATION

There are a number of countries, which cover their electricity demand predominantly from hydroelectric power production (Table 4.4). This makes these countries very vulnerable to droughts and other natural phenomena which hinder a stable and secure electricity supply. These countries need an electricity source diversification.

While attempting to diversify away from an over-reliance on hydroelectric power in order to improve energy security, many countries increased the share of high-emission fossil fuel power plants, which often required increased fossil fuel imports. An alternative is that these countries consider developing domestic geothermal resources of low- and high-temperature potentials, which are an ideal low-emission option, not dependent on climatic events. This can be observed in the Central American region (especially El Salvador, Honduras and Nicaragua), where because the 1980s droughts affected the security of electricity supply, the share of thermoelectric generation was increased during the 1990s (Bundschuh *et al.* 2007a). This was done in spite of the absence of significant oil and gas fields in the region (Guatemala is the only country with some oil resources). Most of the fossil fuels were imported from Mexico and Venezuela. As a consequence, these Central American countries have become more dependent on imported fossil fuels to support their growing electricity demand, instead of accelerating the development of local available, renewable energy resources. Many countries are following the same path. This is evident from the worldwide renewables share of total electricity generation, which corresponds predominantly to hydroelectric power, and which is expected to decline from 19% in 2004 to 16% in 2030 (see section 2.3.5, Chapter 2).

These countries without domestic fossil fuel resources, as most of the countries listed in Table 4.4 are, can only be recommended to reduce their vulnerability to fluctuations in the world energy markets by developing reliable, environmentally sound, domestic electrical systems from geothermal sources, thereby reducing blackouts and electricity rationing.

Other countries that already have high shares in hydroelectric power (Table 4.4), will continue to develop this source. For example, Guatemala and Costa Rica are two countries that plan to build large hydroelectric projects in spite of some uncertainties regarding the future of hydroelectric power in the region (Table 4.5). There are concerns over the viability of these projects in areas prone to heavy rain and flooding. For example, in September 1999 after two weeks of rainfall and flooding, 100,000 persons had to be evacuated from areas downstream of the El Cajón hydroelectric project in Honduras. On the other hand, period of droughts or low and variable rainfalls can greatly reduce hydroelectric outputs as can be seen in the case of Brazil where power prices escalated during peak times of demand. This calls into question the reliability of hydroelectric power as a continuous, base-load energy source. The susceptibility of these projects to climatic events is one of the dangers of relying too heavily on hydropower supply for the electricity needs of a country or region. In the above example of the El Cajón power plant, which generates 60% of Honduras' electricity, the country has recently suffered from electricity rationing.

Considering these facts, the countries which are already over-dependent on hydroelectric power should reduce their vulnerability to climate-related phenomena, climate events, and natural disasters. It is recommended that these countries develop their geothermal energy resources, which are much more reliable, compared to hydroelectric power, and achieve a well-balanced renewable energy mix between hydroelectric and geothermal sources.

When comparing hydropower with geothermal, it should additionally be considered that the building of hydroelectric dams has been increasingly criticized because of negative effects on the environment and local population. Consideration of environmental and social impacts in the evaluation and planning of hydroelectric projects becomes increasingly important. This is especially true for large hydroelectric projects, which may require the displacement of local populations and the flooding of agricultural areas, forests, and other ecologically sensitive lands. These large hydroelectric projects tend to affect surface and subsurface water flow regimes and quality, and may create or increase water-borne diseases.

The lakes formed behind dams contribute positively to the development of tourism, and irrigation systems, create new employment, and contribute to the development of the infrastructure of remote areas. However, sedimentation, as is the common in all hydroelectric projects, may significantly reduce the lifetime of the venture.

Compared to large hydroelectric projects, geothermal projects have much less environmental and social impacts, and generally are not affected by hurricanes, droughts, flooding, and heavy

Table 4.5. Comparison of potential impacts and risks associated with hydroelectric and geothermal power projects.

	Large hydroelectric projects	Geothermal projects
Critical construction components	Dam, lake	Wellfield, power plant
Vulnerability, risks	Seismotectonic, dam failure, landslides, high climate dependency	Volcanic and seismo-tectonic
Designed lifetime	About 40 years, but possible up to 100 years (or more)	25–40 years or more (under sustainable exploitation);
Environmental and social impacts	Micro- and macroclimate change (evaporation); induced seismicity; large extension; visual impact (positive/negative); flooding of agricultural or natural sensitive areas; relocation of local population; contributes to development of tourism (fishing, boating, windsurfing, recreation) and provides jobs for local and national population	Microclimate change (heat emission into atmosphere); air quality change (H_2S, CO_2, steam emissions into atmosphere); induced seismicity; visual impact, small extension (positive)

rains. Additionally, geothermal installations occupy much less area than hydroelectric projects. Because geothermal is a more constant and more reliable energy source than hydroelectric power, it can be used optimally to supply base-load.

It must be noted though, that the use of geothermal energy is not without possible environmental and social impacts. The extraction of geothermal fluids, which often are highly mineralized, may affect soils and shallow aquifers if they are not properly handled at the surface (see e.g., Birkle and Merkel 2000, 2002; Brown 1995). This impact is minimized by re-injecting the cooled geothermal waste waters back into reservoir at depth. Re-injection may produce micro-earthquakes, but recharges the reservoir and reduces ground subsidence.

Since low-temperature resources are widely distributed exploration sites can be selected in such a way that there is no or minimal conflict with the environment. This is a huge benefit *versus* high-enthalpy sites, which are often in environmentally sensitive areas, as e.g., in volcanic zones that may include national parks, and other protected areas of interest to tourists. Also, because low-temperature geothermal plants can be located almost anywhere, they can easily be placed near urban areas or in areas with existing power transmission lines.

4.5   RURAL GEOTHERMAL ELECTRIFICATION USING LOW-ENTHALPY GEOTHERMAL RESOURCES

Geothermal energy can be used in rural electrification by installing small (<5 MW) low-temperature power plants that could help to improve the development of the region. A detailed discussion on this subject is given in Chapter 10. These smaller plants would be an economic alternative to the costly extension of national grids. They could also improve rural micro- or mini-grids, which classically are based on small generators that burn diesel. This type of fuel is expensive and sometimes not available due to a lack of an appropriate infrastructure, especially roads that are impassable during rainy seasons. Both promotion of direct use and rural electrification based on geothermal resources require the implementation of pilot projects to help popularize these opportunities. Geothermal electricity generation in rural and isolated zones reduces the dependence on oil and diesel fuels, which often must be transported over long distances, and sometimes over environmentally sensitive areas like Lake Nicaragua, where currently, fuel is shipped to Ometepe and other lake islands to support diesel generators (Bundschuh *et al.* 2007a). Here, a spill could harm the unique ecosystem of Lake Nicaragua, which is a principal tourist destination and hence important source of income for this poor country.

CHAPTER 5

Geological, geochemical and geophysical characteristics of geothermal fields

"There is reason to be optimistic about geothermal energy. The exciting period is beginning where anomalous sources of heat are treated as systems. To develop geothermal energy as an important resource one must identify anomalous thermal sources and understand their genesis and geometry."

G. Heiken: Handbook of Geothermal Energy, Los Alamos
National Laboratory, 1982.

5.1 GEOLOGICAL AND TECTONIC CHARACTERISTICS

In any geothermal power generation project, whether it is a low- or high-enthalpy system, it is very important to know the geology, structural and tectonic regime of the area, and subsurface characteristics based on surface geophysical methods, as well as geochemical characteristics of the geothermal waters and gases. This pre-drilling stage of investigation helps to locate the exact sites for undertaking deep exploration/production drilling and hence reduces the cost of drilling. The selected examples of geothermal provinces described in this chapter will provide information on a variety of geological and tectonic features that control the circulation of fluids providing heat to the systems and evolution of geothermal waters and gases. The geological, geophysical and geochemical aspects are dealt with in detail.

Geothermal systems can be classified broadly into four groups based on their geological and tectonic association. The four groups are (1) those associated with active volcanism and tectonism (mostly subduction zones): examples of this group are found in New Zealand, Indonesia, Philippines, and North, Central, and South America (2) continental collision zones: an example of this group is the Himalayan geothermal belt, extending along the Indus suture zone between the Eurasian and Indian plate, (3) continental rift systems associated with active volcanism: the geothermal fields in Ethiopia, Eritrea and Kenya fall in this group, and (4) continental rift systems not associated with volcanism: geothermal fields of this kind occur along the west coast of India and central India and Larderello in Italy. There are other geothermal systems that are associated with asthenosphere doming into the lithosphere. The heat source for such geothermal systems comes from the conduction of heat from the asthenosphere due to ongoing processes of thermal equilibrium between the lithosphere and asthenosphere. The geothermal systems from the above groups that evolved within this geological and tectonic frame will help to understand the importance of geological investigation in geothermal exploration.

Besides the large variety of geothermal systems mentioned above, there is another geothermal system that is associated with large sedimentary basins. These are known as "geopressured reservoirs". They are located in large sedimentary basins where sedimentation has taken place rapidly over geological time. Due to a lack of a mechanism for expulsion of fluids present in the sedimentary pores, the pore pressures of such sedimentary rocks become abnormally high and sometimes approach lithostatic pressures. The geothermal gradient of such strata is also high ($\sim$75 °C/km) due to their deep burial at depths varying from 2 to 3 km (Rowley 1982). Such systems are generally associated with oil and gas reservoirs. Excellent examples of such systems exist in the Gulf of Mexico coastal region. The temperatures of the fluids in such geothermal reservoir rocks vary between 140 to 170 °C with pressures greater than 800 bars. The geothermal reservoirs in such systems occupy a very large area providing high volumes of low-enthalpy—high pressure geothermal fluids suitable for power generation.

5.2   GEOTHERMAL SYSTEMS ASSOCIATED WITH ACTIVE VOLCANISM AND TECTONICS

5.2.1   *New Zealand geothermal provinces*

Low-enthalpy geothermal systems are distributed over a large area in New Zealand and include (1) geothermal waters with discharge temperature of 90 °C and less occurring in the North and South islands; (2) high-enthalpy systems (>150 °C) along the margins of the Taupo volcanic zone (TVZ); (3) low-enthalpy systems with discharge temperatures of 120–160 °C, available from abandoned hydrocarbon wells, and (4) geothermal hot water systems heated near the surface. These systems are associated with four major tectonic settings characterized by (1) subduction related volcanism in the TVZ; (2) intraplate volcanism associated with rifts; (3) fault zones within the North island fore-arc; and (4) Alpine fault zone in the South island fore-arc (Reyes and Jongens 2005). New Zealand has an installed geothermal capacity of 308 MW$_e$. A generalized tectonic setting and the distribution of geothermal areas in New Zealand is shown in Figure 5.1.

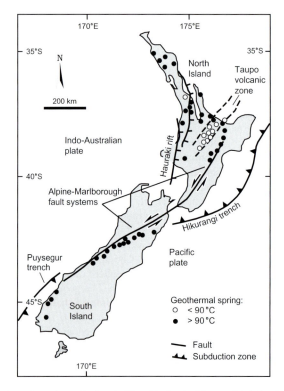

Figure 5.1.   Distribution of geothermal sites in New Zealand. Westward subduction of the Pacific plate and eastward subduction of Indo-Australian plate below the continent has given rise to the Taupo volcanic zone created by the Cenozoic volcanic activity represented by andesitic and rhyolitic volcanism. These two subduction systems are connected by the Alpine-Marlborough fault systems cutting across the North and the South islands. The geothermal fields are located within the Taupo volcanic zone and along the Alpine-Marlborough fault systems (modified after Reyes and Jongens 2005).

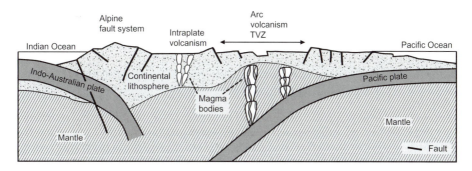

Figure 5.2. Idealized cross section of the tectonic setting of the geothermal systems in New Zealand. The geothermal system is associated with subduction related arc-type and back-arc rift related volcanism. Magmatic intrusions and up-welling mantle provide heat to the geothermal systems. The geothermal waters represent a mixture of meteoric and magmatic waters with volatiles derived from magmatic bodies (modified after Giggenbach *et al.* 1993, Lebrun *et al.* 2000).

The New Zealand microcontinent which is located between the Pacific and Indo-Australian plates, has two boundering subduction zones: one in the east (Hikurangi trench formed due to westward subduction of Pacific plate beneath the continental crust) and a second one in the west (Puysegur trench), enclosing the well known TVZ (Fig. 5.2). The region westward of this subduction is manifested by active andesitic volcanism, crustal thinning, rifting, subsidence, and rhyolitic volcanism resulting in high heat flow (Stern 1987, Gamble *et al.* 1993, Bibby *et al.* 1995). The subduction is further manifested in the Quaternary Taranaki volcano located 140 km SW of TVZ (Downey *et al.* 1994). Cenozoic volcanism is widespread in the South island. The extensional regime that created the Tasman Sea resulted in Early to Middle Tertiary volcanism (Weaver *et al.* 1989). The Miocene volcanism is represented by large intraplate shield volcanoes at Banks peninsula and Dunedin. Pliocene age basalts occur in south Canterbury (Smith 1989).

5.2.2 *Indonesian geothermal provinces*

Indonesian high- and low-enthalpy geothermal systems are distributed around 200 volcanoes located along Sumatra, Java and Bali and along the eastern Indonesian islands (Fig. 5.3).

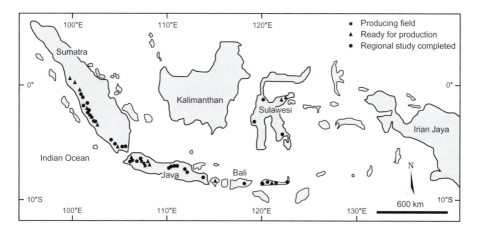

Figure 5.3. Geothermal provinces of Indonesia: All the geothermal provinces are located along the Sumatra fault system (see Figure 5.4) running parallel to the subduction plane and volcanic chain (modified after Fauzil *et al.* 2000).

Sumatra is located on the southern margin of the Eurasian plate and is a part of the Sunda-land craton. The lithology on Sumatra island is predominantly represented by Late Paleozoic meta-sedimentary rocks consisting of limestones, argillites and graywackes. These are the oldest rocks and have been accreted to the Eurasian margin during the Triassic (Stauffer 1983, Cooper *et al.* 1989). The Late Paleozoic sequence is overlain by Jurassic and Cretaceous sediments and mafic volcanics. Late Cretaceous granites occur as intrusions into this sequence (Page *et al.* 1979, Mitchell 1993).

At present, the Indo-Australian plate is moving northwards at 60 to 75 mm/year relative to Eurasia and is undergoing orthogonal convergence below the Java trench (Minster and Jordan 1978, DeMets *et al.* 1990, McCaffrey 1992) (Fig. 5.4). Over the land the Sumatra fault system lies parallel to the subduction. The Sumatra fault system extends NW to the Andaman and Nicobar islands and merges with the Andaman Sea spreading centre (Fig. 5.5) where it acts as a transform fault (Page *et al.* 1979).

Since the Sumatra fault system (SFS) is attached to the Andaman Sea spreading centre, the SFS has acted as a dextral strike-slip fault since the mid-Miocene, the time when seafloor spreading started in the Andaman Sea (Curray *et al.* 1979). The estimated displacement along the fault zone is about 460 km. The fault zone over Sumatra coincides with the volcanic chain represented by Pleistocene to recent volcanoes, thus establishing a close relationship between these two structures (Page *et al.* 1989, McCarthy and Elders 1997). The high-enthalpy geothermal fields are associated with the volcanic areas, and the number of geothermal projects and their resources in the Indonesian islands are given in Table 5.1. All of these high-enthalpy geothermal provinces located within the active volcanic areas are surrounded by large low-enthalpy resources with temperatures measuring ~150 °C. These resources are not yet exploited since high-enthalpy fluids are available in abundance in these regions.

Although the combined geothermal resources are estimated at 27,000 MW (Riki 2005), only 4% of the potential has been tapped until now. Major geothermal sites are located in Java-Bali island and Sarulla of northern Sumatra (340 MW$_e$). Sarulla lies on the NW–SE trending Sumatra fault system that hosts active volcanoes.

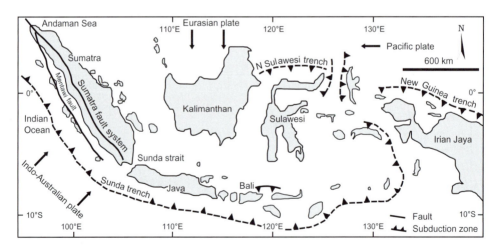

Figure 5.4.   Generalized tectonic map of Indonesian islands (modified after Fauzil *et al.* 2000). The Sunda trech evolved due to the subduction of Indo-Australian plate below Sumatra-Java continent. The Sunda arc continues toward north–west direction giving rise to the Andaman-Nicobar subduction (see Figure 5.5). The geothermal provinces are evolved due to the subduction process and magma generation at shallow levels along the Sumatra fault system (see Figure 5.3).

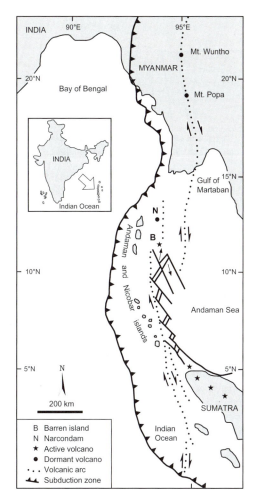

Figure 5.5.   Barren island on Myanmar-Andaman-Java-Sumatra volcanic arc. This volcanic arc has evolved due to the subduction of Indo-Australian plate below the Sumatra island giving rise to the Sunda arc. This arc extends further north where the Indian plate subducts below the Burma (Myanmar) plate. Unlike the Sumatra volcanoes, the Barren island volcano is associated with nascent spreading ridge (modified after Rodolfo 1969).

Table 5.1.   Total high-enthalpy geothermal resources and number of projects developed or under development in Indonesia (source: Fauzi *et al.* 2005).

Island	Number of projects	Total resources MW$_e$
Sumatra	31	9562
Java	22	5681
Sulawesi	6	1565
Others	11	2850
Total	70	19658

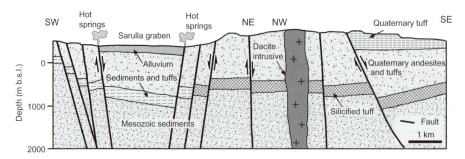

Figure 5.6.   Subsurface geology and structure of Sarulla geothermal field. The sediments and tuff towards the SW side of the area and the silicified tuff extending from the central province to the the SE part of the area act as an insulator containing the heat generated due to magmatic intrusion (modified after Gunderson *et al.* 2000).

5.2.2.1   Sarulla geothermal field

The Sarulla geothermal field is located in the Sarulla graben. The oldest rocks exposed in the subsurface of the graben are of Mesozoic age and are represented by metaquartizites, phyllites, argillites and limestones (Aspden *et al.* 1982), collectively known as the Mesozoic sediments (Fig. 5.6) These Mesozoic sediments are overlain by Quaternary andesites and associated tuffs, which are locally silicified. A dacite intrusive cuts across the entire strata and is exposed towards the eastern side of the Sarulla graben (Fig. 5.6). The entire lithological sequence is traversed by deep-seated faults through which hot fluids ascend to the surface. Three prominent geothermal systems exist within the graben and along SFS with measured bottom hole temperatures varying from 218 to 310°C. The Sibualbuali geothermal system is estimated to generate 40 MW$_e$ for 30 years; the Donatasik geothermal system is estimated at 80 MW$_e$ for 30 years and the Silangkitang geothermal system is estimated to generate a maximum electric power of 210 MW$_e$ in the same time period (Riki 2005).

5.2.3   Philippines geothermal provinces

Major geothermal fields in the Philippines are located in Palinpinon, Bulalo and Leyte (Fig. 5.7). All the fields lie close to the NW–SE trending Philippine fault.

5.2.3.1   Bulalo geothermal field

The Bulalo geothermal field (also known as Mak-Ban field) is located on Luzon island (Fig. 5.7) within the Luzon volcanic arc and has been producing electric power from the geothermal source since 1979. The volcanic arc is characterized by Pliocene to recent volcanic centers. The current production is about 330 MW$_e$, generated from three steam fields, each with individual capacity of 110 MW$_e$, thus saving 40 million barrels of imported oil per year (Clemente and Abrigo 1993). 71 production and 15 injection wells were in operation in 2003. All the wells penetrate to a depth of about 2.8 km to tap the high pressure steam reservoir. Plans are underway to increase the baseload generation capacity to 402 MW$_e$.

The subsurface lithological structure of the Bulalo geothermal field along with the isotherms is shown in Figure 5.8. The geothermal springs emerge from the faults extending to about 2.5 km deep.

The geothermal manifestations in this area are located around several parasitic domes comprising volcanic rocks of silicic composition associated with basaltic cinder cones. The Bulalo dacite dome occurs as an intrusion in the Makling volcanics overlying Makling tuff (Fig. 5.8). A monzonitic intrusion occurs at a depth of about 2000 m. The entire volcanic succession is traversed by a NE-trending fault, exposed towards the eastern side of the dacite dome and two arcuate faults on the western side of the dacite dome (Fig. 5.8). The lowermost succession consists of andesite flows, breccias and tuffs (Pre-Makling series). A monzonite is intruded into the older andesite and tuff (Clemente and Abrigo 1993).

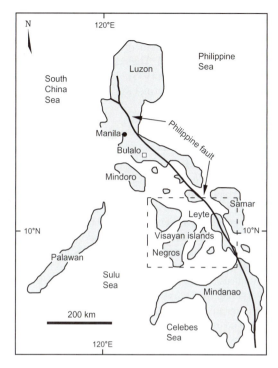

Figure 5.7. Map showing major geothermal fields in the Philippines. All the geothermal fields are located along the main Philippine fault (modified after Rae *et al.* 2004).

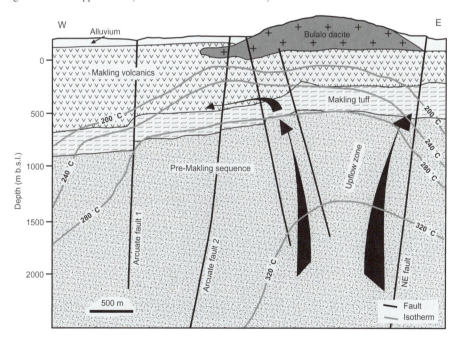

Figure 5.8. Subsurface geology and structure of the Bulalo geothermal province. The monzonite intrusion lying at a depth >2000 m is the main source of heat for this geothermal system. The Pre-Makling tuff is the main geothermal reservoir insulated above by the Makling volcanic rocks. Deep circulating meteoric waters boils at about 1000 m depth and ascends along the major faults (modified after Clemente and Abrigo 1993).

The main upflow zone lies between the NE fault and the Arcuate fault 2. The Makling volcanics form a good insulator and channels the geothermal fluids to flow along the contact between the tuff and volcanic flows. The upwelling fluids emerge along the faults extending to the surface on the western side of the dacite dome (Fig. 5.8). The reservoir temperature measured from the bore wells is around 250 °C (Clemente and Abrigo 1993). As indicated by the isothermal contours, near surface (~500 m depth) low-enthalpy geothermal systems occur within the Makling tuff around the periphery of the Bulalo dacite dome.

5.2.3.2   *Leyte geothermal field*
The Leyte geothermal field is located in the southern part of Leyte island in the Philippines (Fig. 5.7) falling within the domain of Mts. Cabalian and Cantoyocdoc Quaternary stratovolcanoes. These volcanoes are aligned parallel to the regional Philippine fault.

The southern Leyte geothermal field lies over Quaternary volcanics comprising andesites, tuffs and breccias. The Quaternary volcanics lie over Tertiary clastics. Both rock units are intruded by microdiorite and dacite composite intrusives (Fig. 5.9). The Quaternary volcanics vary in thickness from 550 to over 1000 m. The Tertiary clastics include fossiliferous limestones, calcareous sandstones, breccias, siltstones, and claystones. Cretaceous ultramafic rocks lie below the Tertiary clastics, which vary in thickness from ~1200 to ~1700 m. Geothermal manifestation is due to the presence of two prominent intrusions into the entire rock sequence. The Quaternary volcanic flows form an impermeable layer directing the upflowing fluids laterally. Deep-seated faults channel the fluids to the surface. The temperature of the fluids at 1840 m depth is about 280 °C. Exploratory drilling has been completed and the field is ready for exploitation to produce electric power (Rosell and Zaide-Delfin 2005). As evident from the subsurface isotherms, the Quaternary volcanics host low-enthalpy fluids around the periphery of the intrusion over a depth range of 500 to 800 m (Fig. 5.9).

The power generation capacity of the Leyte geothermal field is about 723 MW$_e$. The Tongonan field alone generates 142 MW$_e$ from binary cycle. The power plants are operated by Indonesian power companies under BOT system (Build-Operate and Transfer). Nearly sixteen power plants have been in operation in this region since 1998 (Benito *et al.* 2005).

5.2.3.3   *Palinpinon geothermal field*
The Palinpinon geothermal field is located on Negros island (Fig. 5.7). This field is a high-temperature liquid-dominated system. Bore hole temperature measurements recorded

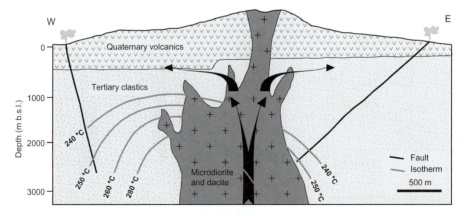

Figure 5.9.   Subsurface geology of the Leyte geothermal field. The Quaternary volcanics include andesite lavas over the surface followed by dacite and andesitic lavas at the lower part of the volcanics. The Tertiary clastics include limestones, breccias, sandstones, and claystones. Microdioritic intrusives provide the heat for the geothermal systems (modified after Rosell and Zaide-Delfin 2005).

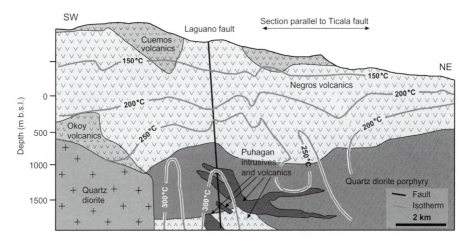

Figure 5.10. Subsurface geology of the Palinpinon geothermal field. The isotherms are drawn based on borehole temperature measurements. The plane of intersection of the Ticala and Lagunao faults is the main upflow zone of the geothermal fluids and also defines the zone of large intrusives, like the Puhagan volcanics and diorites, providing heat to the geothermal system. Lateral migration of the geothermal fluids along these faults is indicated by the lower temperatures isotherms (modified after Rae *et al.* 2004).

temperatures of 225 °C at a depth of about 1 km. The Negros island is a part of the Visayas region and lies over the back-arc of the Negros trench. The current tectonic activity results in horst and graben structures in this region (Rae *et al.* 2004). The Palinpinon geothermal field is hosted by Pleistocene volcanic rocks of Cuernos de Negros volcano (Pornuevo 1984). The subsurface lithology of this region seen along the SW–NE trending Ticala fault is shown in Figure 5.10. The Cuemos volcanics, comprising andesites and dacites, overlie the Late Pliocene to Early Pleistocene Negros volcanics represented by hornblende andesites and andesitic volcanoclastic rocks. A small part of Oky volcanics lay below the Negros formation on the SW part of the section. The entire Negros volcanics have a large areal extent and have a thickness of about 1000 m. The 150 °C isotherm is located at a shallow depth within this rock unit, making the area excellent for future development of low-enthalpy geothermal resources. The base of the sequence is represented by diorite intrusives. The Puhagan dikes and intrusives appear to be the youngest intrusives in the diorites. The Laguano and Ticala faults are the most prominent deep-seated faults that appeared to have localized the Puhagan magma emplacement as well as up-flowing geothermal fluids in this region (Rae *et al.* 2004). The combined electric power generation from the Palinpinon geothermal fields is about 193 MW$_e$ (installed). 43 production and 26 re-injection wells are in operation. The Palinpinon I field generates the maximum power of 113 MW$_e$ while the Palinpinon II field, consisting of Balas-Balas, Nasuji and Sogongon power plants, generate 20, 20 and 40 MW$_e$, respectively (Benito *et al.* 2005). All the above power plants are in operation since 1977.

5.2.4 *Central American geothermal provinces*

Central America is characterized principally by the convergent boundary where the Cocos and Nazca plates are subducting beneath the Caribbean plate. This plate collision creates crustal instability and results in the 1100 km long Quaternary volcanic chain extending from the Mexico-Guatemala border to central Costa Rica, and after a gap in southern Costa Rica continues in Panama (Fig. 5.11). As a result, the region is endowed with rich geothermal resources whose distribution is principally along the Quaternary volcanic chain, shown in Figures 5.11 and 5.12. A detailed overview of geothermal resources in Central America and the potential regional benefits of development is given in Birkle and Bundschuh (2007a, b) and in Bundschuh *et al.* (2007a). This section briefly describes a few examples of geological and tectonic features

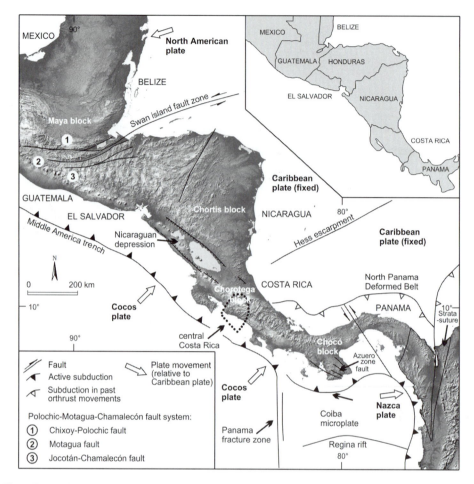

Figure 5.11.   Geo-tectonic setting of the Central American region (modified after Bundschuh *et al.* 2007b).

associated with the geothermal areas to show the huge potentials of low-enthalpy resources available in the Central American region and to highlight their benefits compared to high-enthalpy resources.

According to the estimates of Gawell *et al.* (1999), the geothermal provinces in all these countries have a potential to generate 13,210 MW$_e$, which must be considered as a lower estimate as explained in Chapter 3. National estimates are more conservative: about 1000 MW in Guatemala, 100 MW in Honduras, 400 MW in El Salvador, 1100 MW in Nicaragua, and 900 MW in Costa Rica, totalling 3500 MW (no data available for Panama) (Bundschuh *et al.* 2007a). However, only a small percent of this potential is being exploited at present for power generation covering about 7% of the regional power production (Bundschuh *et al.* 2007a) and direct applications. The countries which were producing power in 2007 were Guatemala (Zunil and Amatitlán), El Salvador (Berlín and Ahuachapán), Nicaragua (Momotombo), and Costa Rica (Miravalles). The total installed capacity in Central America region was 434 MW$_e$ and power generation was 2595 GWh in 2005, which demonstrates that only a small part of the confirmed resources has been used for power generation. Additionally, these countries, consider only high-temperature geothermal resources for power generation and neglect the potential of low-enthalpy resources, which are much larger and more widely distributed compared to high-temperature resources, which are limited to small areas. Costa Rica is the first country in Central America, which through the support

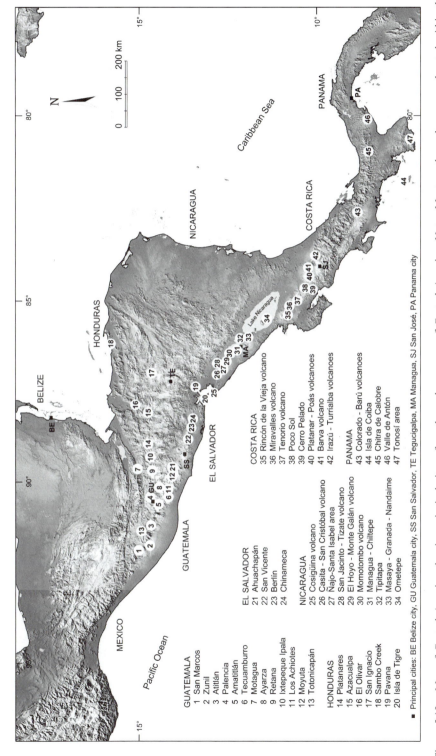

GUATEMALA
1 San Marcos
2 Zunil
3 Atitlán
4 Palencia
5 Amatitlán
6 Tecuamburro
7 Motagua
8 Ayarza
9 Retana
10 Ixtepeque Ipala
11 Los Achiotes
12 Moyuta
13 Totonicapán

HONDURAS
14 Platanares
15 Azacualpa
16 El Olivar
17 San Ignacio
18 Sambo Creek
19 Pavana
20 Isla de Tigre

EL SALVADOR
21 Ahuachapán
22 San Vicente
23 Berlín
24 Chinameca

NICARAGUA
25 Cosigüina volcano
26 Casita - San Cristóbal volcano
27 Najo-Santa Isabel area
28 San Jacinto - Tizate volcano
29 El Hoyo - Monte Galán volcano
30 Momotombo volcano
31 Managua - Chiltepe
32 Tipitapa
33 Masaya - Granada - Nandaime
34 Ometepe

COSTA RICA
35 Rincón de la Vieja volcano
36 Miravalles volcano
37 Tenorio volcano
38 Poco Sol
39 Cerro Pelado
40 Platanar - Poás volcanoes
41 Barva volcano
42 Irazú - Turrialba volcanoes

PANAMA
43 Colorado - Barú volcanoes
44 Isla de Coiba
45 Chitra de Calobre
46 Valle de Antón
47 Tonosí area

■ Principal cities: BE Belize city, GU Guatemala city, SS San Salvador, TE Tegucigalpa, MA Managua, SJ San José, PA Panama city

Figure 5.12. Map of Central America showing principal sites of geothermal resources of Central America. Note: Most geothermal areas are located within the volcanic chain; only the geothermal prospects of Honduras derive from deep-circulation systems related to extensional faulting structures (after Birkle and Bundschuh 2007a).

of the German government, has started to evaluate its low-enthalpy resources. Other countries, like Honduras and Panama which performed detailed reconnaissance studies of their geothermal resources during the past 3 decades, abandoned their geothermal programs since the temperatures of the reservoir fluids, in most cases, were significantly below 200 °C, and hence were considered not viable for commercial power generation. This was true some years ago before new technologies for improved heat exchangers became available. However, with the current available drilling and binary technologies it is possible to exploit these low-enthalpy resources for power generation (Birkle and Bundschuh 2007a). For these two countries the low-enthalpy option is an ideal solution, whereas the countries which have high-enthalpy resources need to additionally consider the exploitation of their low-enthalpy resources that surround large areas of the high-temperature resources.

The wide distribution of available low-enthalpy resources compared to isolated areas with high-temperature resources not only increases the electricity generation potential, but additionally allows a much more flexible site selection for geothermal plants. This becomes especially evident if we take into account that practically all the high-enthalpy resources are located in volcanic zones that are environmentally sensitive areas, including national parks and other protected tourist areas (Bundschuh *et al.* 2007a). This is especially true in Costa Rica, where protected areas cover about 25% of the land (i.e., nearly its entire main mountain range which comprises volcanoes hosting geothermal high-enthalpy sites). The Rincón de la Vieja and Tenorio volcanoes are two examples. The Rincón de la Vieja site, is the most promising geothermal prospect. However, most of its high-enthalpy geothermal resources are located within the national park, which limits the exploration and exploitation activities to areas outside the boundaries of the national park. Such problems can be avoided, if low-enthalpy reservoirs are considered for development.

5.2.4.1   *Guatemala*
In Guatemala, exploration for geothermal resources was started in the 1970s. In spite of the country's vast geothermal resources, only 3% of the national power is being tapped from this source (Roldán 2005). Moyuta, Zunil, Amatitlán, and Totonicapán are geothermal sites with temperatures ranging from 230 to 300 °C, and were evaluated as the best prospects, followed by San Marcos and Tecuamburro. The Los Achiotes, Moyuta and Ixtepeque-Ipala sites were evaluated as lower potential zones while Palencia, Retana, Ayarza, Atitlán and Motagua sites were identified as lowest potential zones (Lima *et al.* 2003, Palma and García 2003, Caideco 1995) (Fig. 5.12). The two geothermal provinces, which are presently used for power generation are Amatitlán, located 30 km south of the national capital Guatemala city, and Zunil, about 200 km west of Guatemala city. The subsurface geological section of Amatitlán geothermal field is shown in Figure 5.13. The granitic basement is about 500 m deep and is overlain by volcanic succession of different ages. The thick volcanics form excellent insulation, creating possible sites for obtaining geothermal fluids with temperatures of ~150 °C, suitable for developing low-enthalpy geothermal power projects.

5.2.4.2   *Honduras*
In contrast to most other geothermal areas in Central America, the geothermal provinces in Honduras derive heat not due to Quaternary volcanism, but from deep-circulation systems related to extensional fault structures. Platanares, San Ignacio, Azacualpa Sambo creek, Pavana, and El Olivar are the main geothermal provinces in Honduras where exploration started in the 1970s (for locations see Figure 5.12). The Platanares field, located in the western part of Honduras, represents the best high-enthalpy geothermal site with measured temperatures of about 160 °C at the shallow depth of 250 m (Sussman 1995), and reservoir temperature was estimated to be around 225 °C.

The remaining geothermal sites provide mainly low-enthalpy fluids with estimated reservoir temperatures varying from 139 to 180 °C (Fig. 5.14), which have hindered their exploitation until now. These sites need to be re-examined in the light of newly available advanced drilling and heat exchanger technologies for power production using the binary technique.

Figure 5.14 shows the subsurface geological strata of the Platanares geothermal province. The geothermal site is enclosed by a NW trending graben. The La Bufa fault located towards the northern

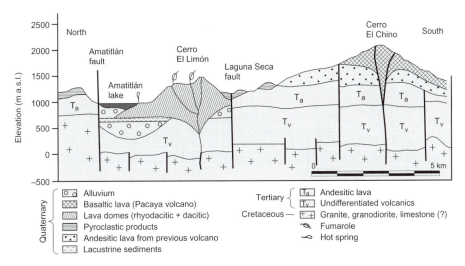

Figure 5.13. Geological cross-section of Amatitlán geothermal reservoir (modified from Birkle and Bundschuh 2007a based on a compilation of data from Roldán 1992, Tobias 1987).

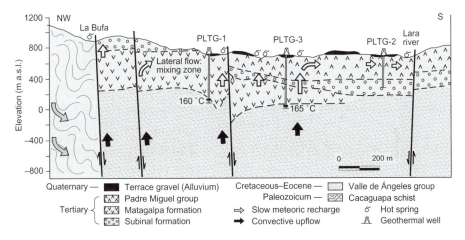

Figure 5.14. Conceptual model of Platanares geothermal reservoir site with principal faults, rock units and recharge and discharge areas (modified after Goff *et al.* 1991).

part of the graben controls major geothermal discharge sites. The volcanic sequence within the graben lies over Paleozoic metamorphic basement rocks. The volcanic rocks include Tertiary andesites overlain by volcanic tuffs. Sedimentary rocks below the Tertiary andesites indicate their sub-aerial eruption. The contact zone between the older metamorphics and younger volcanics is the main recharge site for meteoric-geothermal water circulation (Fig. 5.14).

5.2.4.3 *El Salvador*

El Salvador, like Costa Rica, is one of the countries in Central America that was a pioneer in developing geothermal resources in the 1960s. At present geothermal energy supplies 22% of this country's electricity from the Ahuachapán and Berlín geothermal power plants (151 MW$_e$ installed capacity; 966 GWh power generation).

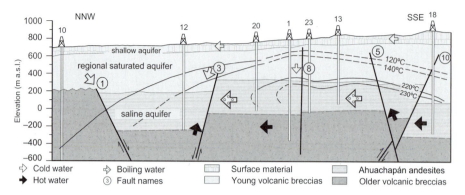

Figure 5.15.   Geological-structural cross-section of the Ahuachapán geothermal reservoir with principal faults (given as numbers), temperature isochrones, location of reservoir, modeled water flow directon, and principal lithological units (modified from Birkle and Bundschuh 2007a based on a compilation of data from Montalvo *et al.* 1997, Aunzo *et al.* 1991, Ripperda *et al.* 1990 and 1991, Laky *et al.* 1989).

The Berlin, Ahuachapán, San Vicente, Chipilapa and Cuyanausul fields are the most prominent geothermal systems in El Salvador. Ahuachapán, located 80 km W of the capital San Salvador, is a liquid-dominated system related to the Ahuachapán-Cerro Blanco graben, while the Chipilapa geothermal system situated about 3 km from Ahuachapán, is related to the Cuyanausul-El Tortuguero graben (see Figure 5.12). These two fields are separated by a triangular uplifted Chipilapa block (González *et al.* 1997). Four principal structural systems that are recognized in this area are: (1) the NW–SE tilted blocks, (2) the NW–SE trending Molina system extending to the andesitic basement, (3) the NNW–SSE trending Cuyanausul system, and (4) the youngest N–S system transecting the Ipala graben.

The volcanic stratigraphy of the Ahuachapán geothermal field is defined by surface lavas and tuffs, young volcanic breccias, Ahuachapán andesites, and older breccias (Fig. 5.15). The entire volcanic succession represents three phases of volcanic activity. The basaltic-andesitic lavas of Cuyanausul, Apaneca and Empalizada were erupted during the Early to Middle Pleistocene ($\sim$1.7 Ma) during the pre-caldera phase. The Concepción de Ataco caldera was formed due to a violent eruption of pyroclastic flows. The post-caldera phase is characterized by Holocene pyroclastic products with 25 m of thickness, called "Zebra pyroclastics". This phase is represented by about 100,000 year old dacitic and andesitic domes.

The younger volcanic breccia is about 400 m thick and forms a top unconfined aquifer traversed by the 120 °C isotherm (Fig. 5.15). The presently existing bore holes drilled to about 900 m depth penetrate the older volcanic sequence with a bottom hole temperature of about 220 °C. As of the year 2000, 16 production wells were drilled producing nearly 95 MW$_e$. The utilized geothermal fluids are re-injected to sustain required pressure in the aquifer and maintain constant flow rate. Re-injection and over exploitation of the reservoirs are a cause of concern since a temperature drop of about 30 °C from 230 °C of the reservoir was noticed, and as a consequence production was reduced.

5.2.4.4 *Nicaragua*

The western part of Nicaragua has high geothermal potential zones. Exploration for geothermal resources in this part of the country began as early as 1966. Since then different geothermal fields were identified, of which Momotombo and San Jacinto are high-enthalpy areas with temperatures above 230 °C. El Hoyo-Monte Galán, Managua-Chiltepe and Masaya-Granada-Nandaime are also highly promising fields with estimated high-enthalpy potentials of 200 MW$_e$ each (Zúñiga Mayorga 2005). Managua and Masaya-Tipitapa were evaluated as high priority areas, Zapatera island as a medium priority, and Ometepe island as low priority prospect (Martínez Tiffer *et al.* 1988) (for locations see Figure 5.12). At present about 271 GWh of power is being generated from Momotombo field, with an installed capacity of 77 MW$_e$, providing nearly 10% of Nicaragua's power.

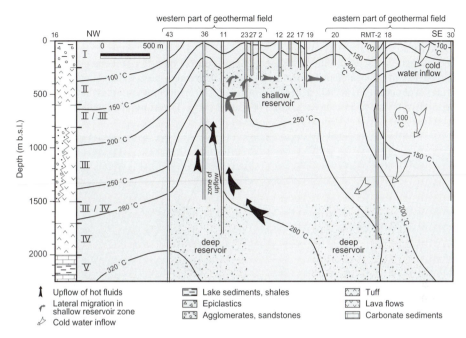

Figure 5.16. Conceptual model of the Momotombo geothermal reservoir, with isotherms, location of reservoir, and lithological column (simplified) (modified from Birkle and Bundschuh 2007b based on compilation of data from Porras *et al.* 2005, Cordon 1980, Goldsmith 1980, California Energy Company, Inc. 1979, Electroconsult 1977, Einarsson 1977, UNDP 1980, Texas Instruments 1971).

The Momotombo geothermal field is located within the rift valley between the Pacific coast and the spurs of the Nicaraguan central plateau. The Nicaraguan depression (Fig. 5.11), occupied by Managua and Nicaragua lakes, was formed due to the outpouring of Quaternary volcanics from the Los Marrabios cordillera formed during the late stages of rift faulting.

The lithology of the Momotombo geothermal field is divided into six units (Martínez Tiffer *et al.* 1988, Porras *et al.* 2005) (Fig. 5.16). The youngest andesites and basaltic andesites interlayered with scoria are underlain by a Pleistocene palagonite tuff alternating with basaltic andesites followed by Late Miocene volcanic products intercalated with sandstones and agglomerates. The middle Miocene fossiliferous marl sediments are separated from the Late Miocene sandstone by ignimbrite flows. The oldest rocks are presented by andesitic lava flows and clastic tuffs. The 150 °C isotherm, located at a shallow depth, occupies a wide area (Fig. 5.16) and makes this field most suitable for exploiting low-enthalpy fluids.

5.2.4.5 *Costa Rica*
5.2.4.5.1 Geothermal development
The first study of the geothermal resources of Costa Rica was initiated in 1961. Within the Guanacaste cordillera, reconnaissance and pre-feasibility studies of the Tenorio, Rincón de la Vieja, and Miravalles geothermal provinces were completed in 1976 (for locations see Figure 5.12). Nine exploratory bore wells were drilled in Miravalles geothermal field between 1979 and 1986.

The national electricity company, ICE, conducted exploratory investigations between 1989 and 1991 and estimated a high-enthalpy potential of 900 MW$_e$ from Costa Rican geothermal provinces. Thus Miravalles, Tenorio, and Rincón de la Vieja geothermal fields were evaluated as having very high geothermal potential; Platanar-Poás was considered to have high–medium potential; Poco Sol, Cerro Pelado, and Irazú-Turrialba are considered as middle potential, and the Barva volcanic area has only a low geothermal potential (for locations see Figure 5.12).

First commercial production of geothermal power started in 1994 with a 55 MW$_e$ power plant. At present this field has an installed capacity of 163 MW$_e$, producing 15% of the electricity generated by Costa Rica.

The evaluations focused only on high-enthalpy resources and do not include the low-enthalpy resources, like those located at Tenorio field, where exploratory bore wells indicate reservoir temperatures of about 160 °C at 2472 m depth (Moya *et al.* 2002) forming resources that can be exploited using modern binary technology e.g. ORC or Kalina cycle. Therefore, the low-enthalpy resources

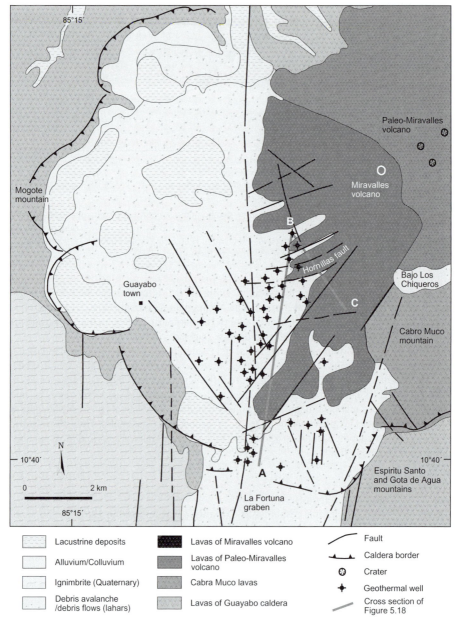

Figure 5.17.   Geological map of Guayabo caldera with Miravalles geothermal field (modified from Birkle and Bundschuh 2007a compiled from Vega *et al.* 2005).

require reassessment in the light of currently available advanced drilling and heat exchanger technologies.

In 2001, the Poco Sol geothermal field was investigated and evaluated as a high-ranking prospect with estimated reservoir temperatures of about 240 °C and a potential of 186 MW$_e$, which is similar to that of the Miravalles reservoir (Vargas 2002). In the framework of the geothermal development program of the German government (Geotherm program), the ICE is planning to conduct a prefeasibility study of the Poco Sol site. This is a sensitive area located near a national park. Taking this into account, the study is focused on utilizing binary power generation technology in developing this site.

In the 2001–2004 period, a feasibility study was conducted for the Las Pailas geothermal province, which is located in the Rincón de la Vieja volcanic area. Five exploration wells were drilled to depths varying from 1418 to 1827 m. The bottom-hole temperatures recorded were between 229 and 244 °C. One well recorded a temperature of 168 °C (Moya 2006) making it suitable for installing a binary power plant if permeability is suitable.

5.2.4.5.2 Miravalles geothermal field

The Miravalles area, which is the best studied geothermal system in Costa Rica, is located in the NW–SE trending Quaternary volcanic chain of Guanacaste, where the regional stress system resulted in a series of NW–SE, NE–SW, and N–S trending fault systems. The geothermal reservoir is characterized by an active hydrothermal system within the 15 km wide caldera type structure (Fig. 5.17). The top of the liquid-dominated reservoir, which has a temperature of 230–255 °C is located at a depth of about 700 m below ground level.

The geology of Miravalles is shown in Figure 5.17 (Birkle and Bundschuh 2007a) and the subsurface lithology and structure of Miravalles geothermal field is shown in Figure 5.18. Water-rock interaction processes on core samples obtained from 53 deep wells were studied to understand the subsurface temperature distribution based on alteration products (Vega *et al.* 2005). Three temperature zones are deciphered from the alteration products: a smectite zone indicating a temperature of <160 °C; a transition zone with temperatures between 140 and 220 °C, and an illite zone indicating a temperature of >220 °C (Vega *et al.* 2005). The presence of a high-temperature zone was further confirmed from drill-hole data (see Figure 5.18).

The heat source for the geothermal systems are the Paleo-Miravalles and Miravalles volcanoes (Fig. 5.17). A schematic model developed for the Miravalles geothermal systems is shown in Figure 5.19.

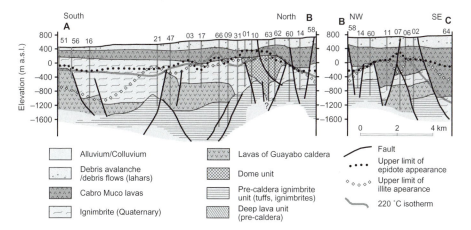

Figure 5.18. Geological cross sections and temperature distributions obtained from measurements in wells and from clay mineral distributions through Miravalles geothermal well field (wells are indicated by numbers; for locations of cross-section see Figure 5.17) (modified from Birkle and Bundschuh 2007a compiled from Vega *et al.* 2005).

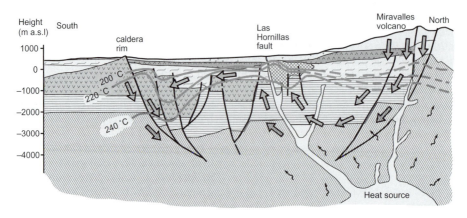

Figure 5.19.   Conceptual model of Miravalles geothermal reservoir. The principal heat source is supposed to be at Miravalles and Paleo-Miravalles volcanoes. Recharge of fluids through faults is predominantly along the border of the Guayabo caldera and the hills containing Miravalles and Paleo-Miravalles volcanoes. Upflow is along the faults within the caldera, (e.g., Las Hornillas fault). Location of the cross-section is shown in Figure 5.17 (modified from Birkle and Bundschuh 2007a as compiled from Vega *et al.* 2005). For legend see Figures 5.18.

5.2.4.6   *Panama*

In Panama, geothermal exploration started in the 1970s resulting in several nation-wide inventories of major hot spring areas (with maximum spring water temperature of 72 °C) as potential geothermal areas comprising Barú-Colorado volcanic complex, Valle de Antón, Tonosí area, Coiba island, and Chitra de Calobre (for locations see Fig. 5.12). The drilling of 6 exploratory wells to depth of 949 m at the Barú-Colorado volcanic complex in 1976/77 indicated only moderate temperature gradients of less than 90 °C/km.

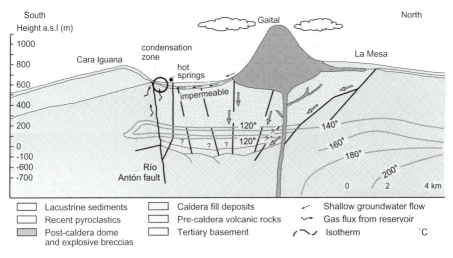

Figure 5.20.   Geological and gravimetric interpretation of El Valle de Antón geothermal area located about 80 km west of Panama city in a complex Quaternary volcanic structure underlain by the Tertiary volcanic basement of the Panamanian mountain chain. The subsurface temperatures were determined by several geothermometers (modified from Bundschuh *et al.* 2002 and Birkle and Bundschuh 2007a).

Although geothermal exploration activity was started on the Chitra de Calobre and El Valle de Antón sites in the 1980s, it was suspended in 1992 due to government policy and environmental concern (Bundschuh *et al.* 2002, Birkle and Bundschuh 2007a). In 2000 and 2003 an attempt was made to reinvestigate the geothermal sites around Barú-Colorado, Chitra-Calobre and Tonosi by intiating geophysical and isotopic surveys. However, the exploration program was stopped again, and the geothermal exploration program in Panama remains suspended today.

Based on the existing information on the Barú-Colorado province, where a geothermal gradient of 90 °C/km was recorded, and at Valle de Antón geothermal field, where the reservoir temperature was estimated to be between 140 and 180 °C (Fig. 5.20) the geological and tectonic settings of these geothermal sites appear to bear promising low-enthalpy sites. Other provinces in Panama need to be reassessed for geothermal production potential considering state-of-the-art heat exchanger technology.

5.3 GEOTHERMAL SYSTEMS ASSOCIATED WITH CONTINENTAL COLLISION ZONES

5.3.1 *Himalayan geothermal system*

The Himalayas represent one the best areas of geothermal systems associated with the continent-continent collision zone. The boundary of the Himalayan geothermal belt extending from NW India to NE India with respect to the major tectonic features is shown in Figure 5.21.

The Himalayas are one of the most complicated but interesting geological and tectonic regimes represented by rocks from the Precambrian to the recent. As a result of collision between the Indian and Eurasian plates, three important zones that have been developed are the Main Boundary thrust (MBT), the Main Central thrust (MCT), and the Indus suture zone (Fig. 5.21). The Higher Himalayas zone, constituting the main metamorphic belt, overthrust the Lesser Himalayan rocks along the Main Central thrust (MCT). The Higher Himalayas zone has suffered the maximum crustal shortening and represents the region of the highest uplift in the Himalayas. The Higher Himalayas zone exposes a 8–10 km thick slab of metamorphics and granites from a deeper level of the crust to its present position (Thakur 1987). The granites vary in age from 5.3 to 60 Ma

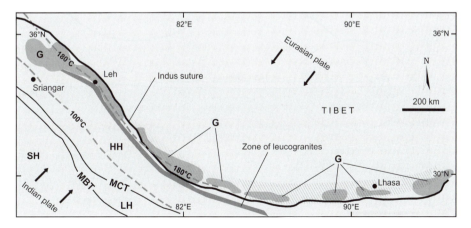

Figure 5.21. General tectonic and geological features of the Himalayas. Hatched area represents approximate boundary of the Himalayan geothermal belt (HGB). SH: Sub-Himalayas; LH: Lower Himalayas and HH: High Himalayas; G: Granite batholiths. Leucogranites are the youngest igneous rocks in this region generated due to shallow (~7 km) crustal melting. The granites have high contents of radioactive elements. Both, shallow magmatic activity and high heat generating capacity of the granites make this region best suited to develop enhanced geothermal system (modified after Chandrasekharam and Chandrasekhar 2007).

(Schneider *et al.* 1999a, b; Searle 1999a, b; Le Fort and Rai 1999; Haris *et al.* 2000; Harrison *et al.* 1998, 1999). These are the younger granites exposed all along the Indus suture zone (Fig. 5.21). The oldest granites of Permian age (268 Ma) occur in the northwestern part of Himalayas (Zanskar region); anatectic processes leading to the intrusion of granites of 1 Ma have also been reported in Nanga Parbat, in the western part of High Himalayas (Schneider *et al.* 1999c). International Deep Profiling of Tibet and the Himalayas (INDEPTH) project located seismic bright spots in the Tibet region (east of the Indian geothermal provinces), which are attributed to the presence of magmatic melts and/or saline fluids within the crust (Makovsky and Klemperer 1999). Highly saline fluids are also found in Ladakh granites (~60 Ma) as inclusions which are attributed to the high volatile content in the granitic melts (Sachan 1996). These crustal melts appear to have been generated within 7 km of the continental crust within the Indus suture zone. A schematic cross section across the Himalayas showing the main structural and thermal regime is shown in Figure 5.22. The leucogranites that are wide spread all along the Indus suture zone (Fig. 5.21) appear to have been generated through anatectic melting and are the main source of geothermal manifestations all along the Himalayan geothermal belt (Hochstein and Regenauer-Lieb 1998). The ^{3}He/^{4}He ratios in geothermal gases in certain parts of the Tibet geothermal provinces indicate that the heat generated from radioactive elements is the main source causing melting of the shallow crustal rocks within the Indus suture zone (Kearey and Wei HongBing 1993, Hochstein and Regenauer-Lieb 1998, Hoke *et al.* 2000). Both subduction tectonism and shallow crustal melting are the main causes for the present day observed heat flow values ($>$100 mW/m^2; Fig. 5.21) all along the geothermal belt.

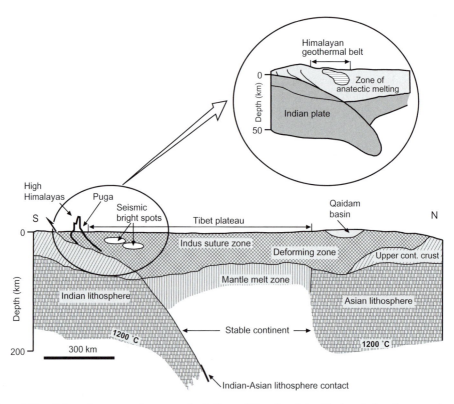

Figure 5.22.   Schematic cross section across the Indian and Eurasian plates. The insert gives the approximate location of the granitic melts. The zones of shallow crustal melting have been recognized through INDEPTH (see text). The Puga, Nagqu and Yangbajing geothermal fields are located within the Tibet plateau (modified after Hochstein and Regenauer-Lieb 1998 and after Hoke *et al.* 2000).

The Himalayan geothermal belt (HGB), located at an altitude of >4000 m a.s.l., extends over a length of about 1500 km from NW to SE of India (Fig. 5.21) and hosts more than 100 geothermal springs. The most prominent geothermal province of India is located in Puga, within the Ladakh province (Fig. 5.21), while in China the most important geothermal field is located in Yangbajing province. At Puga both water and steam-dominated systems exist. The high heat flow ($>100\,mW/m^2$) in HGB, as described above, is due to the presence of a large number of relatively young granite intrusives and shallow crustal melting processes. These granitic intrusives lie below a thick sedimentary cover and are under compressive stress regime (Chandrasekharam 2001b, c). The surface temperature of the geothermal discharges at Puga is around 87 °C. The entire Puga valley is covered with borax deposits and in some places with sulfur deposits. The Himalayan geothermal discharges, in general, are saline in nature indicating mixing of saline magmatic fluids (Alam *et al.* 2004).

5.3.1.1 *Yangbajing geothermal field, China*
The geothermal fields in China are distributed along the east coast and along the Indus suture zone within the HGB (Fig. 5.23). The Yangbajing geothermal area is located in the Cenozoic basin bordered by the Nyainquentanglha mountains on the northwest and the Tang mountain chain on the south.

The Yangbajing geothermal field is located 94 km northwest of Lhasa, at an elevation of about 5000 m a.s.l. It is a high-temperature geothermal field located on the eastern side of the HGB (Fig. 5.21). The mean annual air temperature and pressure in this region are about 2.5 °C and 600 millibar, respectively. This field falls within the Cenozoic basin that is aligned parallel to the Indus suture zone (Dor Ji and Zhao Ping 2000). The basin is bordered by the Nyainquentanglha mountains on the northwest and the Tang mountains on the south. The Nyainquentanglha mountains host paragneisses intercalated with granitic gneiss and amphibolites. The region is traversed by two prominent faults that trend NE–SW and NW–SE. The geothermal system is controlled by the NW–SE trending faults. The Yangbajing field is divided into the northern and southern parts by the nearly E–W running Nepal-China highway. The topographically lower southern part is covered by

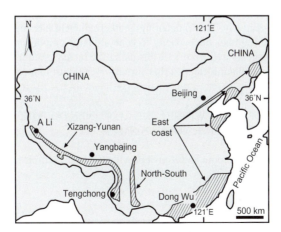

Figure 5.23.   Geothermal provinces in China. Geothermal provinces are shown in hatched shade. The Xizang-Yunan, Yangbajing an Tengchong geothermal provinces, with heat flow values varying from 91 to 146 mW/m^2, are part of the Himalayan geothermal belt shown in Figure 5.22. These three provinces enclose nearly 400 geothermal springs. The low-enthalpy geothermal fields are located along the east coast provinces. They enclose about 500 hot springs with surface temperatures of 60 °C. The north-south fields enclose about 100 geothermal springs with temperatures varying from 60 to 90 °C (modified after Kearey and HongBing 1993).

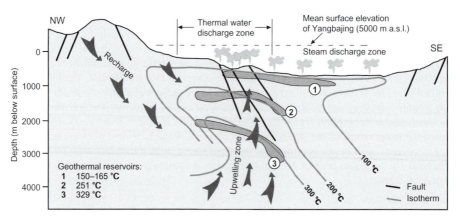

Figure 5.24.   Schematic cross section of Yangbajing geothermal field. 1: Shallow reservoir; 2 and 3: Deep fractured granite reservoir. This field falls within the HGB (see Figure 5.22). Production wells tapping the zones 1 and 2 are generating 28 MW$_e$ (modified after Dor Ji and Zhao Ping 2000).

100–300 m thick Quaternary alluvium underlain by Himalayan granite and tuff. Opal and calcite precipitate from the geothermal waters and form a self-sealed alluvial cap. Altered granite with a small amount of volcanic breccia underlies the northern part. The granite is often replaced by kaolinite, and drilling cores of well ZK4002 give a K-Ar age of 7.6 Ma. The volcanic breccia has a K-Ar age of 46 Ma, much older than the granite. The granite is fractured at depth in the northern part of the field. Pervasive sericite and chlorite alteration is present in mylonite gneiss with biotite also being chloritized. A magnetotelluric survey, performed by UN/DDSMS in 1995, revealed a resistivity anomaly at a depth of 5 km in the northern part of the field, which is inferred to be a magmatic body (Zhao *et al.* 1998a).

A schematic cross section of the Yangbajing geothermal field is shown in Figure 5.24. Fractured granites are the main reservoirs in this area. Geophysical survey and exploratory bore well logging located three reservoirs in this field. A shallow reservoir, hosted in alluvium and weathered granite, located at a depth of about 200 to 300 m has fluid temperatures of 150 to 165 °C. Two deeper reservoirs, one extending between 900 and 1350 m and one beyond 1800 m, have fluid temperatures of 251 and 329 °C, respectively. The Yangbajing geothermal field has been producing electricity since 1970. Though the field has the capacity to generate large amounts of electricity, at present only 28 MW$_e$ are being generated from 13 wells using the binary system (Bertani 2005). This is the only geothermal power plant located at this altitude (~5000 m) in the world.

5.4   GEOTHERMAL SYSTEMS WITHIN THE CONTINENTAL RIFT SYSTEMS ASSOCIATED WITH ACTIVE VOLCANISM

The geothermal systems in Ethiopia and Kenya provide the best examples of this type.

5.4.1   *Ethiopian geothermal fields*

Geothermal manifestations in Ethiopia are distributed within the Ethiopian rift valley and are located in Dallol, Tendaho rift, Aluto Langano, Corbetti, and Abaya (Fig. 5.25). Aluto Langano and Tendaho rift are the two most important provinces. Aluto Langano is located in the rift valley, 200 km SE of Addis Ababa, while the Tandaho rift is located 600 km NE of Addis Ababa (Fig. 5.25). Exploratory wells drilled in the Aluto volcanic complex in the Aluto Lagano province indicate temperatures in the order of 180 °C at 1800 m depth, and reaching 335 °C at a depth of 2200 m (Endeshaw 1988).

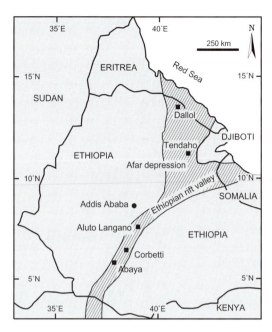

Figure 5.25. Map showing the geothermal provinces in Ethiopia (modified from Endeshaw 1988). All the geothermal provinces are located within the Afar depression and Ethiopian rift valley.

The Tendaho rift is an important geothermal province extending across about 2500 km^2. Geological, gravity, geoelectric, and magnetometric surveys indicate the existence of a reservoir with temperatures above 200 °C in Dubti, north of the Tendaho rift. An exploratory bore well in Dubti confirmed the existence of a liquid-dominated shallow reservoir at a depth of about 500 m. The reservoir is hosted in the permeable zone of lacustrine sedimentary sequence. A basaltic lava flow below this sedimentary sequence has a low permeable zone. However, the Dubti fault, travers- ing the Dubti volcanic flows, shows upflow of geothermal fluids that feed the shallow geothermal reservoir. The geothermal fluid temperature along the faults is about 245 °C.

The Dubti geothermal field is located in the northern Tandaho rift, which is a part of an active NW–SE trending rift basin filled with lacustrine deposits and post stratoid basaltic flow. The lacustrine deposits include siltstones, sandstones, and a small portion of mudstones. At some sites the sedimentary sequence is interlayered by recent basalt sills that increase in frequency with depths below 600 m. The lacustrine sedimentary sequence is underlied by basaltic flows referred to as the Afar Stratoid series. The Afar Stratoid series includes basalt flows interlaid by ignimbrites and detritic deposits that represent the floor of the Tendaho sedimentary basin. The Tendaho rift is about 50 km wide, bordered by escarpments composed of Afar basalts, and tectonically characterized by open fissures and active normal faults that divide the basin into linear NW–SE blocks. The Dubti fault that is a part of this major fault system is 2 km long and is the locus of surface geothermal manifestation, consisting of isolated boiling mud craters, fumaroles, steaming grounds, and hydrothermal deposits (Battistellia *et al.* 2002).

The Aluto-Langano geothermal field has been exploited since 1999, generating 7.5 MW$_e$ from a binary plant. The field has the potential of 30 MW$_e$ (Bertani 2005). Exploratory investigations in the Dubti geothermal field have been completed and the field is expected to generate 3.5 MW$_e$ from its shallow reservoir (Battistellia *et al.* 2002).

5.4.2 *Kenya geothermal fields*

The Kenya rift valley is a foci intense rift related to volcanism and also hosts active geothermal systems. Kenya's geothermal resources are controlled by two major tectonic regimes, one associated

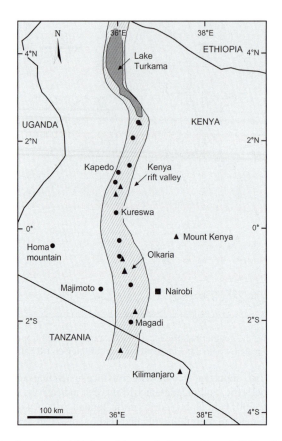

Figure 5.26. Map showing principal high- and low-enthalpy geothermal fields of Kenya. The rift valley encloses all the major volcanoes. High-enthalpy geothermal fields (e.g., Olkaria, see text) are located within the rift valley and low-enthalpy geothermal fields are located along the flanks of the rift valley and surrounding isolated volcanic areas. Triangles: Quaternary volcanoes; Solid circles: geothermal fields (modified after Riaroh and Okoth 1994).

with Quaternary volcanoes, and the other associated with fissures related to an active fault system along the flanks of the rift. Those that are associated with Quaternary volcanism are all high-enthalpy systems, while the low-enthalpy systems are prevalent along the flanks of the rift valley (Fig. 5.26).

5.4.2.1 *Olkaria geothermal field*
The Olkaria high-enthalpy geothermal field had started generating 15 MW$_e$ of power in 1981, and at present is one of the major geothermal power producing fields of Kenya with the production of 127 MW$_e$ (Mwangi 2005). The Olkaria plants 1 and 2 are single flash power plants (5 units), while Olkaria 3 (2 units) and Oserian are binary plants. The Oserian binary plant was installed in 2004 and is generating 4 MW$_e$.

The Olkaria volcanic complex consists of about 80 volcanic centers. Both central and fissure eruptive styles of volcanism have given rise to pyroclastic cones and thick lava flows (Clarke *et al.* 1990). The volcanic rocks include peralkaline rhyolites. A simplified volcanic stratigraphy of Olkaria geothermal field is shown in Figure 5.27.

The Olkaria geothermal field is represented by a succession of basaltic and rhyolitic volcanic rocks capped by recent pyroclastics and interlayered by tuffs. The Pleistocene magmatic activity

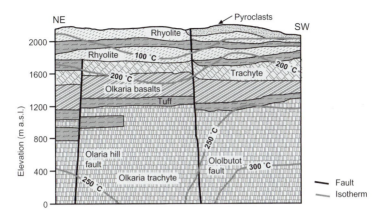

Figure 5.27. Simplified volcanic stratigraphy of Olkaria (NE) geothermal field. Deep-seated fault systems located over an active volcanic region provide heat to the deep circulating meteoric waters. The lateral migration of the geothermal fluids is represented by near horizontal isotherms (modified after Omenda 1998).

resulted in a large eruption of trachyte flows. Minor flows of basaltic and rhyolitic composition also erupted during the Pleistocene. The rhyolites represent the youngest eruption (post Pleistocene) associated with pyroclastics (Fig. 5.27). The trachytic and basaltic flows are separated by layers of tuff of similar age (Omenda 1998).

5.4.2.2 *Low-enthalpy geothermal fields*

Several low-enthalpy systems occur along the flanks of the rift systems (Fig. 5.26) and can be exploited for generating power (Tole 1988). The surface temperature of springs varies from 52 to 95 °C. They include those issuing at the Homa mountains, Majimoto, Kapedo, Kureswa, and Magadi (Fig. 5.26).

Homa mountain springs: These springs with surface discharge temperature of 64–90 °C, occur near Homa bay. The springs discharge through fenitized carbonatites covered by lacustrine sediments. Quartz geothermometry gave reservoir temperatures of about 142–179 °C.

Majimoto springs: These springs emerge near the town of Narok, through the Mozambiquian metamorphic and Tertiary volcanic rocks. Surface discharge temperatures vary from 52 to 57 °C and quartz geothermometry gave a value of 92 °C for the reservoir.

Kapedo springs: These springs occur near Kapedo with a surface discharge temperature of 52 °C. Quartz geothermometry gave a value of 126 °C for the reservoir. The springs flow through a volcanic ridge and traverse Pleistocene sediments.

Kureswa springs: These springs emerge along the Kerio valley which is a part of the Kenya rift valley. The springs flow through Miocene phonolites and have a surface discharge temperature of 63 °C. The reservoir temperature estimated through quartz geothermometry gave a value of 122 °C.

Magadi springs: The banks of Magadi lake, located in the southern part of the Kenya rift valley, hosts nearly 200 geothermal springs with maximum surface discharge temperatures of 95 °C. Holocene sediments are exposed over the entire area and are underlain by Pleistocene trachytes and by a Pliocene alkali basalt suite that forms the base of the basin; this is underlain by the Archean basement. The entire basin is crossed by several faults that are part of the main rift system. Quartz geothermometry indicates the reservoir temperature to be 150 °C.

5.5 GEOTHERMAL SYSTEMS ASSOCIATED WITH CONTINENTAL RIFTS

Low- and high-enthalpy geothermal systems associated with major continental rifts are best represented by those occurring in Larderello, Italy and those found in the Indian subcontinent.

5.5.1   *Larderello geothermal field, Italy*

The geothermal field at Larderello is one of the few systems producing superheated steam. It is now generally accepted that the superheated steam at Larderello is produced by the massive transfer of heat from the rocks to the fluid (Minissale 1991). With the exception of two wells in the southern part of the field that discharge a two-phase fluid, all the others produce superheated steam. There is no evidence from the discharge of the wells, of liquid water being present in the reservoir. A mass balance based on the total cumulative production of the field (26 Mt/year since at least 1951; Chierici 1961) however, suggests that the steam produced at Larderello must be stored in the reservoir as water (Marconcini *et al.* 1977), unless the steam reservoir extends below the Moho, as provokingly reported by James (1968). In their conceptual model of steam-dominated systems, White *et al.* (1971) suggested that water remained trapped in the rock matrix and in the smaller fractures, even though steam was the pressure-controlling phase in the larger fractures.

The Larderello geothermal field is located in the pre-Apennine belt of southern Tuscany. This area has been characterized by extensional tectonics since the Late Miocene. Sporadic compressive stages occurred in the Pliocene. The result of these tectonic phases is a block-faulting structure with NW–SE trending horsts and basins. Small post-orogenic granitic stocks were emplaced along the main axes of the uplifted structures. The anomalous heat flow that marks the Larderello area is likely to be related to one of these intrusions (Minissale 1991). A schematic section of the Larderello geothermal field is shown in Figure 5.28.

The reservoir, comprising Middle Triassic to Oligocene carbonates (limestones), overlying the Paleozoic metamorphic rocks (phyllites, mica schists and gneiss) holds the superheated steam derived from deep circulating meteoric water at 250 °C and above. The carbonate reservoir is overlain by Eocene, Miocene and recent pelitic sediments (Cataldi *et al.* 1963).

Larderello geothermal field was the first in the world to generate electric power. Prince Piero Ginori Conti conducted an experiment using the geothermal steam to generate electric power in 1904. Using a heat exchanger, Prince Conti heated a secondary fluid to produce high pressure vapor and ran a piston engine fitted with a 10 kW dynamo. The electric power thus generated from the dry steam was able to light 5 bulbs. This was the first binary cycle plant that produced electric power from geothermal source. This success was followed by a 3.5 MW$_e$ power plant in 1923 and a 12.15 MW$_e$ plant in 1930 (Lund 2004). This was escalated to 132 MW$_e$ in 1943. The entire plant was destroyed during the World War II in 1944. However, the field was re-established in 1950 to generate 300 MW$_e$ of electric power. The present generation capacity of the Larderello dry steam fields is about 790 MW$_e$ (Lund 2004).

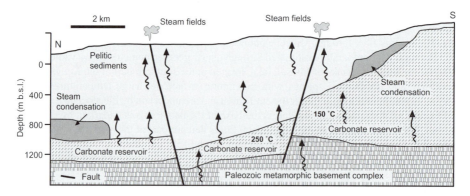

Figure 5.28.   Schematic cross section of Larderello geothermal field. Carbonate rocks form the main reservoir in this field that stores both steam and hot water. Dry CO$_2$ vents are common in this field (modified after Minissale 1991).

Several low-enthalpy geothermal resources occur around the high-enthalpy systems in Italy and around the volcanic islands near Sicily. At present these systems are being used for direct applications.

5.5.2 *Low-enthalpy systems of India*

Low-enthalpy geothermal systems associated with major rifts occur (1) along the west coast of India, (2) along the Son-Narmada-Tapi (SONATA) mid-continental rift system in central India, (3) within the Godavari and Mahanadi graben, and (4) within the Cambay graben in Gujarat and Rajasthan (Fig. 5.29). These provinces are characterized by geothermal springs with issuing temperatures varying from 47 to 98 °C. The reservoir temperatures estimated using chemical geothermometry vary from 100 to 200 °C.

5.5.2.1 *West coast geothermal province*

All eighteen geothermal springs located along the west coast issue through 65 Ma Deccan flood basalts (DFB) with surface temperatures varying from 47 to 72 °C. The thickness of the basalt flows varies from 2500 to 3000 m along the coast and these lava flows are traversed by a large number of N–S trending faults and dyke swarms (Hooper 1990). The major tectonic feature along the west coast is the west coast fault, which controls all the geothermal springs (Chandrasekharam 1985, Minissale *et al.* 2000). The DFB province is seismically active with frequent occurrence of low to moderate earthquakes of magnitude 3.5–6.0 (Chadha 1992). The geothermal gradient recorded from boreholes is about 57 °C/km (Chandrasekharam 2000). The lithosphere along this province is about 18 km thick with 1250 °C isotherm located at shallow depth of 20 km (Pande *et al.* 1984). Presence of such high thermal regime at shallow depth results in high heat flow value (75–129 mW/m^2).

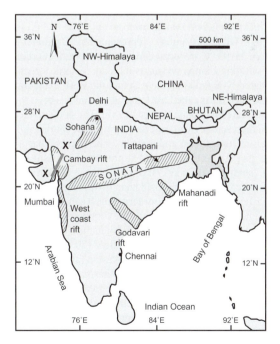

Figure 5.29.   Map showing the geothermal provinces associated with continental rift systems. The Cambay and SONATA geothermal provinces are associated with continental rift tectonics. The Tattapani geothermal field (see text) is associated with high heat generating granites (modified after Chandrasekharam 2000). X–X′ indicates the cross-section shown in Figure 5.31.

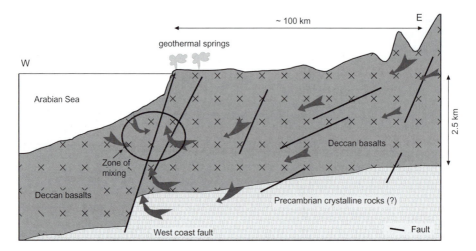

Figure 5.30.   Geothermal fluid circulation along the west coast of India. The geothermal system is generated due to thin continental crust along the west coast of India and the location of a 1250 °C isotherm at shallow depth. Saline water mixing occurs during the ascent of the geothermal fluids along the west coast fault (modified after Minissale *et al.* 2000).

The reservoir temperatures estimated from gas and cation geothermometry of the west coast sub-provinces vary from 120 to 150 °C. The mechanism of evolution of the geothermal manifestations along the west coast as proposed by Minissale *et al.* (2000) is shown in Figure 5.30.

5.5.2.2   *Gujarat and Rajasthan geothermal provinces*
The geothermal provinces in Gujarat (Cambay), represented by 22 geothermal springs, with issuing temperatures varying from 35–93 °C are located in a wide spectrum of lithology ranging in age from Archean to Quaternary with the Cambay basin as its main tectonic structure. The Cambay basin, bounded by two deep-seated N–S faults on the east and the west, extending down to mantle depth (Kaila *et al.* 1981), was formed during the Late Cretaceous and has been rotated anticlockwise during post northward drift of the Indian plate (Biswas 1987). The 4000 m thick Quaternary sediments enclosed by the basin are represented by marine and continental continental claystones, sandstones, conglomerates, fossiliferous limestones, and gypsum. The Cambay basin contains few structural traps with hydrocarbons that are being exploited at present. The basin is a foci of major alkaline magmatism (Sheth and Chandrsekharam 1977). Granite intrusives such as the 955 Ma Godhra granite (Gopalan *et al.* 1979) outcrop within the basin near Tuwa, where geothermal springs with 93 °C issue (Fig. 5.31). Besides the two major faults bordering the basin, several ENE–WSW fault systems are present in this basin. The SONATA rift axis cuts the Cambay fault system towards the SE part of the rift basin. The depth to Moho varies from 32 to 36 km in the Cambay basin giving rise to positive gravity anomaly of +35 mgals, high geothermal gradient (70 °C/km), and high heat flow values (67–93 mW/m^2). Mantle degassing along these deep-seated faults is indicated by a relatively high R/Ra ratio of 0.3 and high CO_2 contents (3%) in the geothermal gases (Minissale *et al.* 2003). A few bore wells in the Cambay basin yielded steam at depths greater than 1500 m with a discharge rate of 3000 m^3/day (Thussu 2002). The flow rate of geothermal water in certain locations like Dholera and Lasundra is more than 1000 l/s with surface temperatures of 46–53 °C (Minissale *et al.* 2003).

Archean to recent lithological units enclose the geothermal field towards the NE part of Rajasthan (North of Cambay and Sohana; Fig. 5.29). The entire tectonic regime of NE Rajasthan, which hosts several geothermal springs, has developed over paleo-subduction tectonic systems (Sinha Roy *et al.* 1998). The Great Boundary fault demarcates the contact between the post and the Precambrian rocks, represented by the Vindhyans on the east and the Precambrian rocks, represented by the

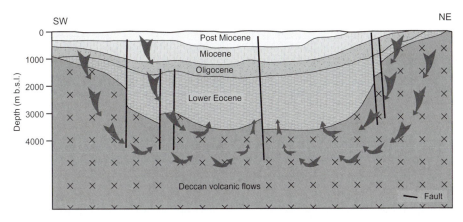

Figure 5.31.   Structure and lithology of Cambay geothermal field. The sedimentary rocks produce oil and gas, and geothermal fluids are encountered in certain oil wells in the basin. The Cambay rift system is part of the west coast fault system reactivated during the Deccan volcanic eruption in the Cretaceous. The Cambay rift is a failed arm of the triple junction. The other two arms are represented by SONATA and the west coast fault (modified after Biswas 1987, Sheth and Chandrasekharam 1997, Minissale *et al.* 2003). For location of cross section see Figure 5.29.

Aravallis on the west. This fault was reactivated during the Eocene due to counterclockwise rotation of the Indian plate and continues to be active at present (Biswas 1987, Minissale *et al.* 2003). Several NE–SW trending fault systems developed due to cyclic and dynamic movement of the blocks and form channels towards deep circulating geothermal waters. The southern part of Rajasthan is covered by fluvial and lacustrine deposits accumulated in the NE–SW trending graben structure developed during the Precambrian (Sinha Roy *et al.* 1998, Minissale *et al.* 2003). The surface temperature of the geothermal waters is about 50 °C.

The reservoir temperatures estimated using chemical geothermometers in waters and gases vary from 120 to 150 °C. These low-enthalpy geothermal provinces are under stages of exploration for producing electrical power using the binary system.

5.5.2.3   *SONATA geothermal province*

The Son-Narmada-Tapi lineament (SONATA) represented by the Narmada-Tapi rift system, is located between the Indo-Gangetic plain in the north and the Precambrian shield in the south. The WSW–ENE trending structure that includes SONATA (Fig. 5.29) is considered to be a mid-continental rift system. This rift system was formed due to the interaction between the two proto-continents (Naqvi *et al.* 1974) during the early stages of the Indian plate development and reactivated after the collision of the Indian plate with the Eurasian plate (Ravisankar 1991, Bhattacharji *et al.* 1994). Deep seismic sounding profiles across the SONATA, south of Tattapani, suggest that the fault system reaches the mantle (Kaila *et al.* 1985). The Tattapani springs (Chandrasekharam and Antu 1995) are located at the eastern edge of the SONATA and are related to the Balarampur fault system. They flow through Archean metamorphic rocks consisting of quartzites, schists, gneisses intruded by granites, pegmatites and amphibolites (Joga Rao *et al.* 1986). The SONATA is a focus of several earthquakes of moderate magnitude and is characterized by high helium (0.54–6.89%) and high CO_2 contents in the geothermal gases (2.88%; Minissale *et al.* 2000). Similarly the Ar content varies from 1.09 to 1.6%. The surface temperatures of the geothermal waters at Tattapani (Fig. 5.29) vary from 30 to 93 °C. A schematic cross-section showing the geology and structure of the Tattapani geothermal system is shown in Figure 5.32. Both steam and hot water issue from shallow bore wells with a flow rate of 50 l/s. The entire SONATA has a sedimentary insulation over granitic intrusive at about 2 km depth (Chandrasekharam and Prasad 1998) producing a geothermal

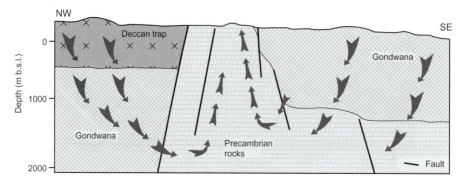

Figure 5.32.   Geology and structure of Tattapani geothermal field. The Tattapani geothermal system is located in high heat producing granites (due to radioactive decay) and issues through the local fault intersections that lie parallel to the main SONATA rift (modified after Chandrasekharam and Antu 1995).

gradient of 60 °C/km. The high geothermal gradient at Tattapani (~90 °C/km) is due to the presence of radioactive minerals in the Precambrian rocks (Chandrasekharam *et al.* 2006), as indicated by the high helium content. The occurrence of shallow granitic intrusives below a thick sedimentary thermal insulation makes SONATA one of the best enhanced geothermal systems in the country.

5.5.3   *Geothermal resources of Mongolia*

Compared to the other geothermal systems discussed above, Mongolia represents an entirely different geothermal regime associated with multiple deformed geological and tectonic events that occurred during pre Mesozoic time. This country has no active volcanism or tectonic activity but still possesses a considerable amount of low-enthalpy resources.

The country is divided into the Central Asian orogenic belt and an accretionary orogen that occurred during the Neoproterozoic to Late Paleozoic arc-system formation (both island arc and continental margin arc system) containing several microcontinents that consist mostly of Archean Proterozoic metamorphic rocks of ~2646 Ma. The most prominent structure is the E–W trending Main Mongolian lineament. This linement separates the northern and the southern terranes. The northern terranes were amalgamated during the Ordovician while the southern terranes were amalgamated during the post Paleozoic time. Central Mongolia experienced multiple intense tectonic and volcanic events during the end of the Mesozoic and the middle Tertiary resulting in mountain building activity that is commonly inferred to have been associated with the South Khangai hot spot activity, itself associated with the evolution of the Khangai mountain ridge. Morphometric analysis of the Mongolian-Siberian mountain belt indicates upwelling of the asthenosphere below this region (Ufimtsev 1990, Zorin *et al.* 1990) giving rise to an anomalous asthenospheric bulge in this region. This asthenospheric bulge results in high heat flow values in central Mongolia (Fig. 5.33). The anomalous hot mantle has not equilibrated with the overlying crust and conduction from this mass is providing the required heat for circulating meteoric water, thereby giving rise to a low-enthalpy reservoir within the entire central Mongolian belt (Doraj 2005, Bignall *et al.* 2005).

About 40 geothermal springs with flow rates varying from 1 to 50 l/s with temperatures varying from 20 to 92 °C are distributed around Khangai and Khentii. High-temperature geothermal springs are found in Khuvsgul, Zavkhan, Arkhangai, Bayankhongor, Uvurkhangai, and Khentii provinces. All these geothermal springs have traditionally been used for physiotherapy, and attempts are now being made to install district heating systems in these provinces (Dorj 2005).

The geochemical data show that all the geothermal springs are rich in bicarbonate (Gendenjamts 2005) indicating either mixing of ascending geothermal water with near surface groundwater or geothermally heated groundwater. The anomalous reservoir temperatures estimated using Na-K-Ca

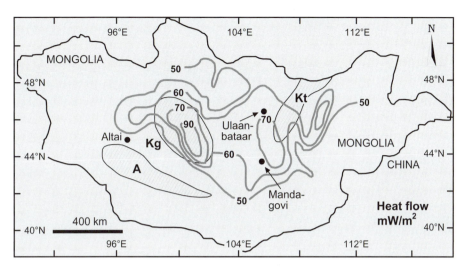

Figure 5.33.   Heat flow contours and geothermal provinces of Mongolia. A: Altai, Kg: Khangai, Kt: Khentii provinces (modified after Doraj 2005 and Lkhagvadorj 2005).

geothermometer gave unrealistic temperatures reiterating the role of near surface groundwater in controlling the composition of the geothermal waters in this region (Gendenjamts 2005). However, the chalcedony geothermometer which gave values varying from 69 to 123 °C appears to be accurate for this region (Gendenjamts 2005).

The geothermal waters from the Khangai region are being used for greenhouse cultivation. Geothermal space heating was initially utilized in 1973 for a health resort and subsequently for a sanatorium and a restaurant using hot water with 89 °C, and a flow rate of 5.5 kg/s (Lkhagvadorj 2005). Large scale space heating systems using geothermal water have been in operation since then. The return water with a temperature of 50 °C is being used for balneology. Thus, geothermal resources are being utilized extensively for direct application in Mongolia.

In Mongolia the urban population receives poor quality of electricity. In 2002 nearly 33% of the urban population did not have access to electricity and 43% of the population had no access to a central heating system. Similarly out of 314 villages, only 117 had access to supply grid based electricity. The remaining villages were using ∼60–100 kW capacity diesel generators for 3 to 5 hours a day (World Bank 2002). The diesel has to be imported and transported for long distances to these rural areas. Due to supply interruptions and poor maintenance, power supply interruptions and frequent breakdowns of equipment are hampering socio-economic development (Bignall *et al.* 2005).

The Khangai province appears to be a promising site for initiating binary power projects using low-enthalpy resources. Exploratory shallow bore wells drilled near Shagajuut in Khangai province, to a depth of about 90 m, yielded geothermal water with a temperature of 48 °C (Fig. 5.33). The geothermal manifestation in this province includes steaming ground with boiling water (98 °C) discharging at the rate of 50 l/s. Assuming a flow rate of 60 t/h and a water temperature of about 120 °C, it is estimated that a binary power plant may be able to generate 300 kW of electricity to meet the demand of about 2500 people living in the Shagaljuut area (Tseesuren 2001, Dorj 2001).

CHAPTER 6

Geochemical methods for geothermal exploration

"The ultimate objective of any exploration programme is to locate a resource that can be economically developed. Despite differences in the type of resources and their geological settings a certain exploration philosophy has been built up over many decades. This philosophy is based on the concept that the prospector begins search in a large area and narrowing down to a more specific location."

A.W. Laughlin: Hand-book of Geothermal Energy,
Los Alamos National Laboratory, 1982.

6.1 GEOCHEMICAL TECHNIQUES

Uprising magma or deep-seated intrusives are the main sources of heat for wet geothermal systems (Fig. 6.1) where the chemical characteristics of the geothermal fluids are controlled by water-rock interactions at high temperatures. Due to involvement of magma intrusion, several chemical constituents present in the magma mix with the circulating geothermal fluids. These chemical components, as discussed in the later part of this Chapter, provide vital clues for assessing the reservoir characteristics, its temperature, and other information that is crucial during pre-drilling exploration stages of geothermal resource development. In the case of low-enthalpy geothermal systems in non-volcanic areas (conductive geothermal systems), both the natural radioactivity of elements like uranium, thorium, and potassium, and the conduction of heat from the mantle to shallower levels along deep continental crust, provide heat (Fig. 6.2). In

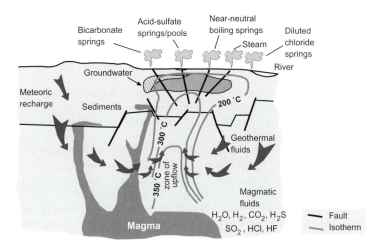

Figure 6.1. Schematic cross section of a typical geothermal system. The heat for the system is provided by magmatic intrusions in volcanic areas. The percolating meteoric waters, indicated by solid arrows, react with the 'host rocks' and mix with magmatic fluids and give rise to different chemical types of geothermal fluids. Ascending geothermal fluids mix with shallow groundwater and change its chemical composition further. The position and lateral extent of the isotherms depends on the volume of the magma chamber or intrusion. Faults provide channels for uprising geothermal fluids (modified after Henley 1985).

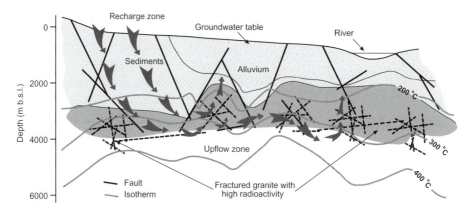

Figure 6.2.　Schematic cross section of a typical geothermal system in a non-volcanic area. The thermal anomaly is generated due to high content of radioactive elements like U, Th, and K in granites. The thick sediment cover over the granites acts as an insulator and prevents upward heat loss. Meteoric waters circulating in such high heat generating granites give rise to geothermal systems. High heat generating granites are excellent sites for enhanced geothermal systems.

both high-enthalpy and low-enthalpy systems, the main source of geothermal fluids is meteoric water. The water-rock interaction processes operate with different degrees in both cases. In the case of geopressured geothermal systems, described in Chapter 5, the chemical characteristics of the geothermal water are controlled by the pore fluids and residence time of the geothermal fluids in the aquifer. The chemical characteristics of the geothermal fluids and geothermal gases, in certain cases, are very important to assess during the exploration and pre-drilling stages. This investigation reduces the number of exploratory and observation wells to be drilled in any geothermal site, "thereby" reducing the over-all cost of geothermal projects.

6.2　CLASSIFICATION OF GEOTHERMAL WATERS

Though the majority of geothermal fluids originate with the percolation of meteoric water, those systems that are derived from magmatic activity contain a sufficient proportion of fluids from other sources. These fluids include juvenile water, magmatic water, connate water, and metamorphic water.

Thus, juvenile water forms a part of the primary magma in the mantle. Deep circulating meteoric water in volcanic terrains that generate high-enthalpy geothermal systems, carry juvenile water originating from such magmas. These fluids can easily be distinguished using the oxygen and hydrogen isotopic signatures of the fluids.

Magmatic waters are those that are associated with magmas. In a sense, they are also meteoric in origin, but later have incorporated into the magma during its formation and emplacement. A good example of such waters are those which are associated with magma generated in subduction zones, where sediments of the subducting slabs carry water that mix with the magma generated.

Thick sediments formed under marine environments contain connate water. These are basically brines commonly associated with oil fields. Such sediments are hot enough to produce geothermal fluids that contain both connate water as well as meteoric water.

Metamorphic reactions also produce water due to the break-down of hydrous minerals. Geothermal systems occurring in such environments have the possibility of incorporating such water. The quantity of metamorphic water when compared to the circulating high volume of meteoric water though, is quite insignificant. However, it is possible to identify such signatures in geothermal systems. Saline fluids entrapped as fluid inclusion in many metamorphic minerals also get mixed with the geothermal fluids.

Table 6.1. Chemical composition of selected typical geothermal waters.

		T	pH	Na$^+$	K$^+$	Ca^{2+}	Mg^{2+}	HCO$_3^-$	SO$_4^{2-}$	Cl$^-$
		°C					mg/l			
NZ	Wairakei W	240	8.5	995	142	17	0.04	5	30	1675
NZ	Wairakei S	99	8	1070	102	26	0.4	76	26	1770
NZ	Broadlands W	260	7.4	675	130	1	0.011	376	41	964
NZ	Broadlands S	98	7.1	860	82	3	0.1	680	100	1060
CR	Miravalles W	250	7.8	1750	216	59	0.1	27	40	2910
CR	Miravalles S	74	8.6	2063	85	33	0.6	739	102	2700
Mx	Cerro Prieto W	280	7.3	5600	1260	333	0.27	40	14	10500
Mx	Cerro Prieto S	80	7.6	5120	664	357	4.0	65	31	8790
Ch	Yangbajing S	70	7.7	425	48	30	0.36	441	27	483
Ch	Yangbajing W	160	8.6	460	64	1	0.14	430	33	550
In	Puga S	86	8.1	611	75	15	1.0	876	70	470
In	Tattapani S	93	8.4	143	7.5	4	0.2	122	38	150
Kn	Homa bay S	87	8.4	7160	241	0.3	0.2	nd	514	2300
Tn	Songwe S	73	6.9	790	102	49	19	1990	160	185
Ep	Tendaho S	95	7.0	506	3.9	24	0.01	33	26	710
Zb	Kapisya S	85	8.6	207	27	10	1	225	268	124
Mn	Khangai S	86	8.8	107	4.2	2.1	0.5	nd	49	17

W: Bore well sample; S: Surface sample. Sources: New Zealand (NZ), Costa Rica (CR), Mexico (Mx): Giggenbach (1988); Peoples Republica of China (Ch): Grimaud *et al.* (1985); India (In): Chandrasekharam, unpublished data, Minissale *et al.* (2000); Kenya (Kn): Tole (1988); Tanzania (Tn): Makundi and Kifua (1985); Ethiopia (Ep): Endeshaw (1988); Zambia (Zb): Sakungo (1988); Mongolia (Mn): Gendenjamts (2005); nd: not reported.

6.3 CHEMICAL CONSTITUENTS IN GEOTHERMAL WATERS

Geothermal waters contain all the major ions found normally in groundwater like Ca^{2+}, Mg^{2+}, Na^+, K^+, HCO_3^-, SO_4^{2-} and Cl^-, but their concentrations in geothermal compared to non-geothermal groundwaters are generally much higher. Similarly the trace element concentration in geothermal waters is far higher than those of the non-thermal groundwater due to its interaction with surrounding rocks at higher temperatures. The geothermal waters however, change their concentration during their ascent to the surface due to several physical processes, like steam loss and mixing with groundwater. Such processes result in different water types at the surface. The Taupo volcanic zone in New Zealand (see Chapter 5.2.1) encloses several geothermal areas and extensive geochemical studies have been carried out on the geothermal water and gases of this geothermal province. These investigations provided vital clues to plan exploration strategies in unexplored geothermal provinces of the world. The geothermal water types described earlier by Ellis and Mahon (1977) are based on the type of geothermal water found and classified in the Taupo geothermal province. In a broader sense, this classification can also be applied to other geothermal provinces of the world associated with volcanism. These water types are shown in the Figure 6.1.

Those fluids that ascend to the surface from the reservoir, without loss of heat (marginal loss of heat due to conductive cooling), will emerge as Na-Cl type water with near neutral pH (Na-Cl; Fig. 6.1). They are found in volcanic settings. These waters have high silica content and the Cl^-/SO_4^{2-} ratio is generally greater than 1 (Table 6.1). Since the geothermal waters emerge directly from the reservoir with high silica content, they are very useful in characterizing the reservoir temperature. Magmatic CO_2 and H_2S are the major gas phases in these waters. Examples of such water types are given in Table 6.1.

Ascending geothermal waters when mixed with the near surface groundwater rich in HCO_3^-, become HCO3-type water with low Cl^- concentrations relative to the Na-Cl waters (Fig. 6.1, Table 6.1).

Geothermal waters with high content of H_2S gas, which condenses near the surface form pools with water of high SO_4^{2-} and low Cl^- contents. Due to oxidation, such pools give rise to low pH and high SO_4^{2-} water at the surface.

Acid-sulfate-chloride water results when oxidation of H_2S, which is of volcanic origin and sometimes contained in large amounts in chloride waters forms bisulfate ions (HSO_4^-). The bisulfate formation increases with decreasing temperature as the water emerges to the surface and results in a low-pH water thereby increasing the Cl^- content. The initial pH of such waters will be neutral at depth and due to the H_2S effect, they become acidic near or at the surface (Fig. 6.1, Table 6.1).

Wall rocks also influence the chemistry of the emerging geothermal waters. Wall rock containing sulfide minerals or sulfide deposits, alter the composition of the water flowing through from neutral to acidic due to the hydrolysis of sulfur. High chlorine gas content in the gas component of the geothermal waters sometimes results in the formation of hydrochloric acid which brings down the pH. These waters show high Cl^- concentrations and low pH values.

6.4 DISSOLVED CONSTITUENTS IN GEOTHERMAL WATERS

6.4.1 *Major ions*

All the geothermal waters carry imprints of deep thermal processes that have occurred at elevated temperatures and pressures within the rock. These waters reside in the reservoir for sufficiently long periods of time (for example thousands of years) so that they are assumed to have achieved chemical equilibrium with the reservoir rock. This is the reason why fast ascending geothermal waters are of significant importance in understanding the reservoir characteristics and provide critical information during exploration. Geothermal waters ascending to the earth's surface either lose or gain chemical constituents depending on the physical and chemical processes that they undergo. The dissolved constituents are a valuable tool during the pre-drilling stage of geothermal exploration. The concentration of solutes in geothermal waters is controlled by mineral-solution equilibria related to the hydrothermal alteration of rocks. Ion exchange reactions involving Na^+, K^+, Ca^{2+}, and Mg^{2+} ions are controlled by temperature and pressure related exchange reactions. For example, K-feldspar alters to muscovite (K-mica) that further alters to kaolinite as the end product:

$$3KAlSi_3O_8 + 2H^+ = KAl_2Si_3O_{10}(OH)_2 + 6SiO_2 + 2K^+ \quad (1)$$
$$\text{(K-feldspar)} \qquad\qquad\qquad \text{(K-mica)}$$

$$KAl_2Si_3O_{10}(OH)_2 + 3H_2O + 2H^+ = 3Al_2Si_3O_5(OH)_4 + 2K^+ \quad (2)$$
$$\text{(K-mica)} \qquad\qquad\qquad\qquad \text{(kaolinite)}$$

During these reactions ions are released into the geothermal waters that carry significant information regarding temperatures under which such reactions have occurred. On the other hand, quartz solubility in geothermal water is controlled by temperature. Thus, silica and cations in geothermal waters provide information regarding temperature and pressure conditions prevailing in the reservoirs.

Based on the thermodynamic properties of the geochemical reactions, the concentrations of major cations like Na^+, K^+, Ca^{2+} and Mg^{2+} vary in the geothermal waters. These cations are termed 'geoindicators' (Giggenbach 1988). They provide information on water-rock equilibrium conditions prevailing in the geothermal reservoirs. The geoindicators are extensively used to estimate the reservoir temperatures in geothermal systems.

As described in section 6.3 the majority of geothermal waters emerging at the surface undergo chemical changes due to various processes. It is necessary to eliminate those altered waters because they are not suitable for estimating reservoir temperatures and take only the 'good' candidates for this purpose. Near neutral chloride type waters are likely to represent well equilibrated fluids from upflow zones (Giggenbach 1988). The geothermal waters with high HCO_3^- and SO_4^{2-}

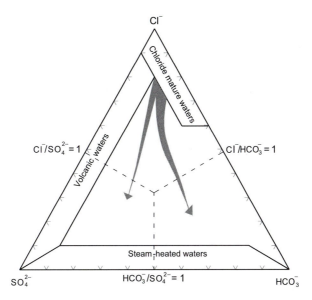

Figure 6.3.   Anion variation in geothermal waters. Fast ascending geothermal waters retain their original chemical composition and fall in the choride mature waters field. The mixing of these geothermal waters with bicarbonate rich shallow groundwater leads to a shift in to the bicarbonate field. Mixing with H_2S gases of volcanic origin makes the waters to shift to SO_4^{2-} field. geothermal springs falling within the chloride mature waters field are good for estimating reservoir temperatures (modified after Giggenbach 1988).

concentrations are not 'good' candidates for assessing reservoir temperatures. These waters can be eliminated by using the three anions *viz*. HCO_3^-, SO_4^{2-} and Cl^- (Fig. 6.3).

Eliminating unsuitable geothermal waters for reservoir chacacterization is essential during the pre-drilling stages of geothermal exploration because the cost of drilling is expensive. Assessing the reservoir characteristics using chemical parameters and geophysical inferences increases the level of confidence.

Fast ascending geothermal waters undergo little chemical change due to water-rock interactions and plot within the chloride mature water field in Figure 6.3. Interaction with near surface water increases the HCO_3^-/Cl^- ratio in the mixed water thereby driving the plot towards the bicarbonate waters field (Fig. 6.3). Similarly, the mixing with high content H_2S volcanic gases will drive the plots towards the sulfate waters field (Fig. 6.3) and render them unsuitable for characterization of the geothermal reservoirs. Those samples falling within the chloride mature waters field are the best suited to characterize the geothermal reservoir, while other samples falling outside (but near to) the chloride mature waters field may be used with a degree of caution. The chemical compositions of geothermal waters from a few wells and geothermal springs from selected geothermal provinces are shown in Table 6.1.

As evident from Table 6.1, the geothermal waters from Broadlands (New Zealand) and Miravalles (Costa Rica) show evidence of mixing with other water sources compared to Cerro Prieto (Mexico) and Wairakei (New Zealand) geothermal waters.

6.4.2 *Silica*

Silica minerals (quartz, cristobalite and amorphous silica) are common alteration/precipitating minerals in geothermal systems. Understanding the behavior of silica minerals is useful to elucidate the reservoir characteristics, temperature of the geothermal reservoir, development of silica alteration zones and to estimate the influence of self-sealing or scaling by reinjection. Silica minerals' solubility is mainly controlled by temperature. Thus, with increase in temperature the solubility of

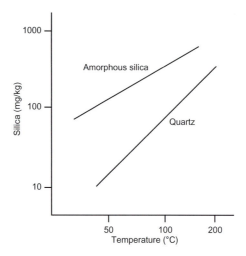

Figure 6.4. Solubility-temperature relationship of silica in geothermal waters. Silica solubility increases with increase in temperature up to 250 °C (modified after Fournier 1973).

silica minerals increases up to about 250 °C. On a temperature *vs* log-concentration plot, dissolved silica in thermal waters lie on a straight line up to 250 °C. This relationship for amorphous silica and quartz is shown in Figure 6.4 (Fournier 1973).

At 25 °C, pressure has less effect on the solubility of quartz and amorphous silica. But at higher temperatures, pressure also influences the solubility of quartz and amorphous silica. Further, at high pressures, variation of temperatures also influences the solubility of quartz and amorphous silica. For example, at 1000 bars the solubility of quartz increases by about 19% at 200 °C and by 36% at 300 °C relative to their solubility at normal pressure (Fournier and Potter 1982). Quartz controls the dissolved silica in all geothermal waters above 180 °C, in most waters with temperatures of 140–180°, and in many with temperatures of 90–140 °C. Calcedony, which is slightly more soluble than quartz controls dissolved silica at temperatures below 91–140 °C, and in some cases at temperatures of up to 180 °C (Fourier 1985). However, the salt content has a considerable effect on the solubility of amorphous silica. When the temperature is constant, solubility decreases with increasing NaCl concentration (Chen and Marshall 1982).

The solubility of quartz in saline solutions for temperatures ranging from 25 to 900 °C can be calculated using the following equation (Fournier 1983):

$$\log m = A + B(-\log \rho F) + C(-\log \rho F)^2 \tag{3}$$

where

$A = -4.66206 + 0.0034063\ T + 2179.7\ T^{-3} - 1.1292 \times 10^6\ T^{-2} + 1.3543 \times 10^8\ T^{-3}$
$B = -0.0014180\ T - 806.9\ T^{-1}$
$C = 3.9465 \times 10^{-4}\ T$
$\rho = $ density of the solution
$F = $ weight fraction of water in the solution
$m = $ molality of dissolved silica
$T = $ temperature in Kelvin

In active geothermal systems hot relatively quickly ascending fluids with temperatures <250 °C precipitate little silica. This phenomenon is of great significance because silica content in hot springs permits the calculation of the reservoir temperatures. Above 250 °C silica minerals precipitate and may seal the channels of flow.

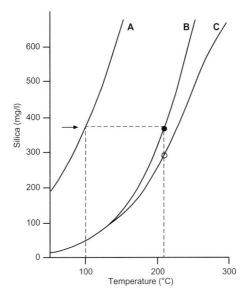

Figure 6.5.   Effect of adiabatic cooling on silica concentrations in geothermal waters. A: Solubility of amorphous silica at vapor pressure; B: Solubility of quartz with maximum steam loss during adiabatic cooling down
to 100 °C given as function of original reservoir temperature. C: Solubility of quartz without steam loss (at
vapor pressure). Open circle: Silica concentration in a 210 °C hot geothermal reservoir in equilibrium with
quartz. Solid circle: Silica concentration in ascending water after adiabatically cooling from 210 to 100 °C
(modified after Fournier 1985).

Adiabatic cooling of rising geothermal waters influences the silica concentration. When fast
ascending water, which was quartz-saturated in the reservoir, cools adiabatically (through boiling
due to lower pressure) without loss of heat to wall rock, and silica does not precipitate, during
ascent, dissolved silica contents in the geothermal waters will increase as steam separates. The
solution will be just saturated with amorphous silica (whose solubility is higher than those of
quartz) at the boiling temperature (100 °C) of the water at the surface (Fig. 6.5). In such cases
corrections can be applied to recalculate the silica in the geothermal waters before the separation
of steam. However, if ascending water undergoes conductive cooling, the silica estimation in the
original water becomes uncertain.

Geothermal waters in the majority of cases, mix with near surface cold water, which causes silica
precipitation. Mixing models have been developed to estimate the silica concentration in original
waters (Fournier 1985, Fournier and Rowe 1966).

For example, let us assume that point 'm' in Figure 6.6a represents local groundwater and
'n' represents geothermal water that mixed with water 'm' and steam did not separate from the
geothermal water before mixing with water 'm'. In such cases, the line joining 'm' and 'n' projected
on the quartz solubility line C (without steam loss) gives the silica content and temperature of the
geothermal water before mixing (point 'o' in Fig. 6.6a).

In case steam is lost before mixing, then the line joining the local groundwater (point 'm') and
the diluted geothermal water (point 'r') (Fig. 6.6b) projected onto the steam loss curve gives the
silica content in the geothermal water after steam loss (point 's'). The silica content before steam
loss can be read by projecting the point s on the silica solubility without steam loss curve (point 't').

As evident from Figure 6.6a and 6.6b, ascending geothermal waters near the surface environment
get supersaturated and result in the precipitation of amorphous silica. Once the silica concentration
in the original geothermal water is established, the values can be utilized to estimate the reservoir
temperature using silica geothermometers.

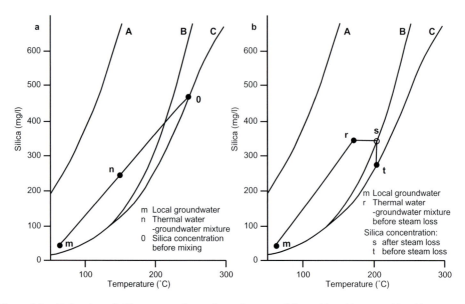

Figure 6.6.   Estimation of silica content in geothermal water mixing with cold water: (a) without steam separation; (b) with steam separation. For other symbols refer to Figure 6.5 (modified after Fournier 1985).

6.4.2.1   *Effect of pH on solubility of silica*

Considerable amounts of CO_2 are always present in steam associated with geothermal waters. Loss of CO_2 accompanied by adiabatic cooling increases pH and so affects silica solubility in geothermal waters. This is due to the dissociation of silicic acid (H_4SiO_4) to $H_3SiO_4^-$, which increases with increasing pH, thereby increasing the silica in the solution. The amount of dissociated H_4SiO_4 can be calculated using the following equations (Fournier 1985):

$$H_4SiO_4 = H_3SiO_4^- + H^+$$
$$K_1 = [H_3SiO_4^-][H^+]/[H_4SiO_4] \tag{4}$$

$$K_1 = (m\ H_3SiO_4^-\ 10^{-pH})/(m\ H_4SiO_4) \times (\gamma H_3SiO_4^-)/(\gamma H_4SiO_4) \tag{5}$$

$$\log m\ H_4SiO_4 = A + B(-\log \rho F) + C(-\log \rho F)^2 \tag{6}$$

$$-\log K_1 = -631.8744 - 0.2967\ T + 0.0000133266\ T^2 + 267.6478\ log\ T$$

$$m\ H_3SiO_4^- = m\ SiO_2(\text{total})/((10^{-pH}\ \gamma H_3SiO_4/K_1) + 1) \tag{7}$$

(parameters in equation 6 are same as in equation 3; T is temperature in Kelvin; activity $= \gamma m$, where m is the molality and γ the activity coefficient).

The effect of pH on the silica solubility is shown in Figure 6.7. Above 130 °C, quartz in contact with solutions of pH > 9.2–9.5 is more soluble than amorphous silica at pH 7 at temperatures between 130 and 300 °C. Therefore, when the pH drops from this alkaline range, where the solubility of quartz exceeds those of amorphous silica to neutral the solution becomes over-saturated with quartz and results in the precipitation of amorphous silica. The occurrence of high pH solutions in the reservoirs is unlikely since phyllosilicate mineral buffering reactions

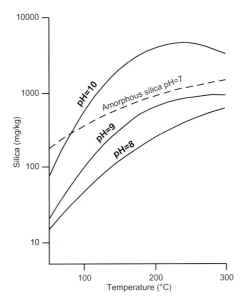

Figure 6.7. Effect of pH on the solubility of quartz between 25 and 300 °C at pH values varying from 8 to 10. The dashed line represents solubility of amorphous silica at pH 7. Above 130 °C and pH > 9.2–9.5, quartz is more soluble than amorphous silica at pH 7 at the same temperature. Alkaline solutions (pH > 9.2–9.5) initially in equilibrium with quartz will precipitate amorphous silica when the pH becomes neutral at a constant temperature (modified after Fournier 1985).

maintain the pH below 7. Even high salinity geothermal waters have pH values below 7. It is unlikely to have a reservoir completely devoid of phyllosilicates. In reactions like:

$$3KAlSi_3O_8 + 2H^+ = KAl_2(AlSi_3O_{10})(OH)_2 + 6SiO_2 + 2K^+$$
$$\text{K-feldspar} \qquad\qquad \text{K-mica} \qquad \text{quartz}$$
(8)

$$K_{eq} = [K^+]^2/[H^+]^2$$
(9)

$$K_{eq} = [K^+]^2 \, [SiO_2]^6/[H^+]^2$$
(10)

precipitation controls the silica concentration in the geothermal waters. The activity of quartz in such cases is unity and the equilibrium constant K_{eq} of this reaction is represented by equation (9).

When silica minerals are absent in the reservoir rock, (i.e. ultramafic/nepheline bearing rocks) then the K_{eq} is defined by equation (10), where a decrease in silica activity leads to pH increases.

In a geothermal reservoir, receiving recharge water from an alkaline source (e.g., saline lakes), the reaction between the water and rocks at high temperature converts the pH buffering minerals to other minerals, such as a feldspar-carbonate assemblage. Such water will have high ^{18}O content indicating the evaporated nature of the water recharging the reservoir.

As mentioned above, CO_2 escaping from the ascending geothermal water can cause an increase in pH in the discharging geothermal waters due to the dissociation of bicarbonate ions (eq. 11):

$$HCO_3^- \rightarrow CO_2 + OH^-$$
(11)

$$3KAlSi_3O_8 + 2H^+ = KAl_2(Si_3O_{10})(OH)_2 + 6SiO_2 + 2K^+$$
$$\text{K-feldspar} \qquad\qquad \text{K-mica} \qquad \text{quartz}$$
(12)

In any geothermal systems, at near surface environment, the rate at which H^+ ions are generated from certain common reactions (e.g., eq. 12) is much slower than the generation of OH^- due to bicarbonate dissociation (eq. 11). Hence, the resulting high pH water tends to dissolve silica.

If quartz is better soluble than amorphous silica, the pH increase may result in a fluid supersaturated with respect to amorphous silica and its precipitation. This results in silica scaling, a major problem in geothermal exploration. Therefore, it is necessary to know the conditions at which amorphous silica precipitates.

The solubility of silica at the vapor pressure of geothermal waters in the temperature range of up to 250 °C can be calculated using the following equation:

$$\log S = 4.52 - (731/T)$$

where T is the temperature in Kelvin, and S is the dissolved silica concentration in mg/kg.

The solubility of molar amorphous silica m at temperatures of 90–340 °C, between vapor pressure and 1000 bars can be calculated using the equations (13) and (14) (Fournier and Marshall 1983):

$$\log m = -6.116 + (0.01625\ T) - (1.758 \times 10^{-5}\ T^2) + (5.257 \times 10^{-9}\ T^3) \tag{13}$$
$$\text{at vapor pressure}$$

$$\log m = -7.010 + (0.02285\ T) - (3.262 \times 10^{-5}\ T^2) + (1.730 \times 10^{-8}\ T^3) \tag{14}$$
$$\text{at 1000 bars}$$

6.4.3 Geothermometers

Estimating reservoir temperatures before any drilling activity is an important task in geothermal exploration. Estimates are based on (1) the solubility of specific minerals like silica minerals, that is controlled primarily by temperature and (2) the exchange reactions of certain minerals that are controlled by temperature and redox conditions. Once the mechanism of silica solubility in geothermal waters, as described above, is well understood in a geothermal system, then the application of silica geothermometers provides valuable information for further exploration of this resource. In the case of exchange reactions, solution-mineral equilibrium conditions have to be understood before estimating the reservoir temperatures.

6.4.3.1 Silica geothermometers

Once the silica content in geothermal water is determined and considering all of the above criteria, it is possible to estimate the correct reservoir temperatures as a prerequisite for the planning of drilling activities. Based on the silica solubility *vs* temperature relationship shown above, equations for the straight lines for the temperature range of 20 to 250 °C have been developed (Fournier 1973), that give the reservoir temperature:

Quartz geothermometer with no steam loss:

$$T(°C) = (1309/5.19 - \log S) - 273.15 \tag{15}$$

Quartz geothermometer with maximum steam loss:

$$T(°C) = (1522/5.75 - \log S) - 273.15 \tag{16}$$

Amorphous silica geothermometer:

$$T(°C) = (731/4.52 - \log S) - 273.15 \tag{17}$$

where S is the silica concentration in mg/l.

6.4.3.2 *Cation geothermometers*

Cation geothermometers are used to estimate reservoir temperatures. These geothermometers are developed based on ion exchange reactions using temperature dependent equilibrium constants K_{eq}. For example, in the reaction involving Na-feldspars (albite) and geothermal water containing K^+ ions (eqs. 18 and 19), the K_{eq} of the reaction can be calculated using equation (20).

In geothermal reservoirs water-rock interactions attain equilibrium conditions due to the prevailing high temperatures and long contact time (residence time) between the rock and the geothermal waters. At high temperatures, temperature-dependent equilibrium exchange reactions are common:

$$NaAlSi_3O_8 + K^+ \rightleftarrows KAlSi_3O_8 + Na^+ \tag{18}$$
$$\text{albite} \qquad\qquad \text{K-feldspar}$$

$$K_{eq} = [KAlSi_3O_8]\,[Na^+]/[NaAlSi_3O_8]\,[K^+] \tag{19}$$

Considering the activities of solids to be unity, equation (19) reduces to equation (20):

$$K_{eq} = [Na^+]/[K^+] \tag{20}$$

For an exchange reaction where monovalent ion like K^+ and divalent ion like Mg^{2+} are involved (e.g. alteration of K-felspar to sericite; see exchange reaction 35), the equilibrium constant can be expressed as:

$$K_{eq} = [K^+]^2\,[Mg^{2+}] \tag{21}$$

where K^+ and Mg^{2+} are molalities of the respective ions.

Similarly, in reactions where water is involved, the K_{eq} is calculated using the following equation:

$$CaAl_2Si_4O_{12} + 2H_2O + 2SiO_2 + 2Na^+ = 2NaAlSi_3O_8 + Ca^{2+} + 2H_2O$$
$$\text{wairakite} \qquad\qquad \text{quartz} \qquad\qquad \text{albite}$$
$$K_{eq} = [Ca^{2+}][H_2O]^2/[Na^+]^2 \tag{22}$$

Hydrolysis reactions are very common in hydrothermal environments and the most affected minerals are potassium and sodium feldspars. Such reactions control the pH of the geothermal waters (eqs. 23 and 24):

$$3KAlSi_3O_8 + 2H^+ = KAl_3Si_3O_{10}(OH)_2 + 6SiO_2 + 2K^+$$
$$\text{K-feldspar} \qquad\qquad \text{K-mica} \qquad\qquad \text{quartz}$$
$$K_{eq} = [K^+]/[H^+] \tag{23}$$

$$2.3NaAlSi_3O_8 + 2H^+ = Na_{0.33}\,Al_{2.33}\,Si_{3.67}\,O_{10}(OH)_2 + 3.33SiO_2 + 2Na^+$$
$$\text{albite} \qquad\qquad \text{Na-montmorillonite} \qquad\qquad \text{quartz}$$
$$K_{eq} = [Na^+]/[H^+] \tag{24}$$

The equilibrium constant can be related to temperature in terms of the van't Hoff equation (25):

$$K_{eq} = C_0 + (\Delta H^\circ/2.303\,RT) \tag{25}$$

where ΔH° is the enthalpy of the reaction, R is the gas constant and T is the temperature in Kelvin and C_0 is a constant of integration.

Since most of the geothermal systems have temperatures of less than 260 °C and ΔH° does not vary much within the temperature range of 0 to 300 °C, equilibrium constant K_{eq} can be substituted by the ratios of activities of various ions to estimate the reservoir temperature (eq. 26).

$$[Na^+]/[K^+] = C_0 + (\Delta H^o/2.303\ RT) \tag{26}$$

Cation geothermometers have been developed based on the above theoretical temperature-related exchange equilibrium reactions. As evident from the above reactions, Na^+/K^+ ratios (as molalities or other units like ppm or mg/kg, etc.) used in the geothermometers may result due to the reaction between albite and hydrogen ions or exchange of Na by K. It is difficult to assess such reactions in natural geothermal systems. Hence, the assumption by various authors results in more than one Na-K geothermometer as given in equations (27–33) where concentrations of Na and K are expressed in mg/kg).

$$T(^\circ C) = 1217/1.483 + \log(Na/K) - 273.15 \tag{27}$$

$$T(^\circ C) = 856/0.857 + \log(Na/K) - 273.15 \tag{28}$$

$$T(^\circ C) = 833/0.780 + \log(Na/K) - 273.15 \tag{29}$$

$$T(^\circ C) = 933/0.993 + \log(Na/K) - 273.15 \quad \text{valid for } (25-250\ ^\circ C) \tag{30}$$

$$T(^\circ C) = 1319/1.699 + \log(Na/K) - 273.15 \quad \text{valid for } (250-350\ ^\circ C) \tag{31}$$

$$T(^\circ C) = 1390/1.75 + \log(Na/K) - 273.15 \tag{32}$$

$$T(^\circ C) = 4410/14.0 + \log(K^2/Mg) - 273.15 \tag{33}$$

(27: Fournier 1983, 28: Truesdell 1976, 29: Tonani 1980, 30, 31: Arnorsson 1983, 32: Giggenbach *et al.* 1983, 33: Giggenbach 1988).

Giggenbach (1988) developed an automatic self-policing method to weed out unsuitable samples (see Figure 6.3) for temperature estimation. This method is based on the reactions shown in equations (34) and (35):

$$\text{K-feldspar} + Na^+ = \text{Na-feldspar} + K^+ \tag{34}$$

$$2.8\ \text{K-feldspar} + 1.6\ \text{water} + Mg^{2+} = 0.8\ \text{Mica} + 0.2\ \text{Chlorite} \\ + 5.4\ \text{Silica} + 2K^+ \tag{35}$$

Thus from equation (34) follows:

$$T(^\circ C_{kn}) = 1390/(1.75 - (\log(K/Na))) \tag{36}$$

where 'kn' represents "K and Na" (K and Na in mg/kg or ppm). From equation (35) follows:

$$T(^\circ C_{km}) = 4410/(14 - (\log(K^2/Mg))) \tag{37}$$

where 'km' represents "K and Mg". K and Mg are in mg/kg or ppm.

The temperature calculated using equations (36) and (37) individually will give different values since the reaction involving K-Na equilibrates at high temperatures while $K/\sqrt{Mg}$ equilibrates at low temperatures. The reaction involving K and Mg equilibrates faster and temperatures estimated from surface geothermal waters give too low reservoir temperatures. Reactions involving K and Na do not adjust quickly to the physical environment at shallow depths. Hence, Giggenbach (1988)

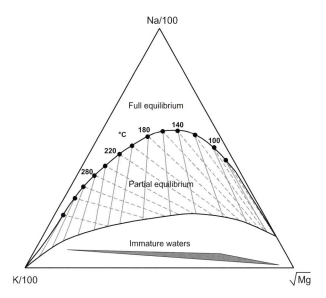

Figure 6.8. Na-K-Mg geothermometers: The lines joining the Na apex and the K-Mg base represent the fast equilibrating K-Mg geothermometer and the lines joining the Mg apex and the K-Na base represent the slow equilibrating Na-K geothermometer. Geothermal water samples falling on the full equilibrium line indicate equilibrium conditions between the geothermal water and host rocks while geothermal water samples falling within the partial equilibrium field indicate mixing of geothermal water with water from other sources while ascending (modified after Giggenbach 1988).

combined both the geothermometers on a triangular diagram (Fig. 6.8) where the concentrations of K-Na-Mg obtained from the geothermal water analyses are plotted to obtain temperatures from both the geothermometers simultaneously.

While plotting the data on this diagram, the concentration of Na^+, K^+ and Mg^{2+} in the geothermal waters (mg/l) is normalized to percentages. The full equilibrium line is drawn based on the assumption that the geothermal water looses steam during ascent.

6.4.4 *Isotopes in geothermal waters*

Isotopic composition of geothermal water, in conjunction with the major ions discussed above, provide information on the evolution of the geothermal water and the processes that affect their composition during water-rock interaction processes. Oxygen and hydrogen isotopes are widely used for this purpose.

6.4.4.1 *Oxygen and hydrogen isotopes in water*
Oxygen has three stable isotopes whose abundance in the crust of the earth are about $^{16}O = 99.74\%$, $^{17}O = 0.05\%$ and $^{18}O = 0.21\%$. Similarly, hydrogen has two stable isotopes whose abundance in the hydrosphere are: $^1H = 99.98\%$ and $^2H = 0.02\%$ (also known as Deuterium, D). Water molecules have nine different isotopic configurations such as $H_2{}^{16}O$, $H_2{}^{17}O$, $H_2{}^{18}O$, $H^2H\,{}^{16}O$, $H^2H\,{}^{17}O$, $H^2H\,{}^{18}O$, $^2H_2\,{}^{16}O$, $^2H_2\,{}^{17}O$, and $^2H_2\,{}^{18}O$.

The oxygen isotopes $H_2{}^{16}O$ and $^2H_2{}^{18}O$ molecules are of immediate concern to geothermal systems. Between these two isotopes, $H_2{}^{16}O$ has higher vapor pressure, while $^2H_2{}^{18}O$ has lower vapor pressure. Hence when water evaporates, vapor gets enriched in ^{16}O while ^{18}O remains in the liquid. Consequently, vapor gets enriched in H while liquid gets enriched in ^2H. The isotopic composition of water is reported with reference to a standard. Standard Mean Ocean Water (SMOW) is considered as the international standard against which all the isotopic ratios of water are reported (Craig 1961).

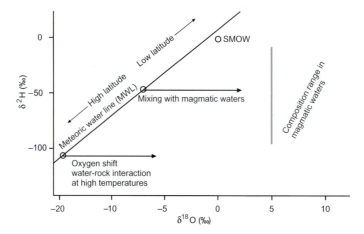

Figure 6.9. $\delta^{18}O$ *vs* δ^2H isotope diagram. Water-rock interactions at temperatures above 200 °C will increase the $\delta^{18}O$ value of the geothermal fluids making the geothermal waters move away from the meteoric line. δ^2H will not show any effect due to its negligible concentration in the rocks (modified after Giggebach and Lyon 1977, Giggenbach and Stewart 1982, Truesdell and Hulston 1980).

The isotopic composition is reported in terms of an expression shown in equations (38) and (39):

$$\delta^{18}O = [(^{18}O/^{16}O)_s - (^{18}O/^{16}O)_{SMOW}]/(^{18}O/^{16}O)_{SMOW}] \times 1000 \tag{38}$$

$$\delta^2H = [(^2H/H)_s - (^2H/H)_{SMOW}]/(^2H/H)_{SMOW}] \times 1000 \tag{39}$$

Equations (40) and (41) are the general expressions for equations (38) and (39):

$$\delta = 10^3[(R_s/R_{SMOW}) - 1] \tag{40}$$

$$10^3 + \delta = 10^3(R_s/R_{SMOW}) \tag{41}$$

where s and SMOW are sample and standard respectively, and R is $(^{18}O/^{16}O)$ or $(^2H/H)$.

Therefore when the water samples are enriched in ^{18}O and ^2H with respect to SMOW, then $\delta^{18}O$ and δ^2H values will be positive. Similarly, when water vapor condenses and falls as rain, the rain will be enriched in ^{18}O and ^2H with respect to cloud thus making the water vapor progressively more negative. The $\delta^{18}O$ and δ^2H of the rain falling will also be negative since the original vapor is depleted in ^{18}O and ^2H.

Physical and chemical processes that affect the geothermal waters can be well understood by plotting the isotopic data on a $\delta^{18}O$ *vs* $\delta^2H(\delta D)$ diagram. One such diagram is shown in Figure 6.9.

Since meteoric water is the main source for all geothermal systems, the $\delta^{18}O$ *vs* δ^2H values of geothermal waters fall on the meteoric water line (MWL). Any deviation from meteoric water line can be explained in terms of exchange of oxygen between rock and water, mixing with magmatic waters, steam loss, etc. (Fig. 6.9). Each case is discussed below.

The position of the local groundwater of any geothermal province on MWL depends on the geographic location of the province. Groundwater from high latitudes and altitudes will plot on the lower end of MWL while groundwater from low latitudes and altitudes will plot on the upper end of MWL.

6.4.4.2   *Oxygen shift*

Oxygen isotope exchange between silicate minerals in the reservoir or wall rocks through which the geothermal waters ascend takes place only at temperatures of 220 °C and above (Henley *et al.* 1985). Since hydrogen content in silicate minerals is negligible, oxygen exchange

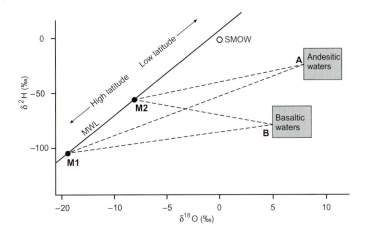

Figure 6.10. Geothermal-magmatic waters mixing relationship. Geothermal waters mixing with fluids from basaltic magmas fall on the M1–B tie line and geothermal waters mixing with fluids from andesitic magmas fall on the M2–A tie line. The boxes define the oxygen and hydrogen isotopes concentration in basaltic and andesitic magmas. MWL: meteoric water line; SMOW: Standard Mean Ocean Water. The oxygen and hydrogen isotope concentrations in geothermal waters tend to be more positive when mixed with magmatic waters (modified after Giggebach and Lyon 1977, Giggenbach 1978, Giggenbach and Stewart 1982, Truesdell and Hulston 1980, and Henley *et al.* 1985).

between silicate minerals and water results in ^{18}O enrichment ($\delta^{18}O$ values become more positive) making the geothermal water samples plot along a line parallel to the $\delta^{18}O$ axis on the $\delta^{18}O$ *vs* $\delta^{2}H$ diagram. This is termed as 'oxygen shift' (Fig. 6.9). The degree of shift is directly proportional to the temperature during the exchange process (Faure 1986). Thus, the farther away the samples plot from the MWL, the more oxygen is incorporated into the geothermal waters.

6.4.4.3 *Mixing with magmatic waters*

In tectonically active regions, like subduction zones, deep circulating geothermal water of meteoric origin may mix with water derived from magmas. The $\delta^{18}O$ of waters derived from basaltic and andesitic magmas range from about 5 to 10‰ (Faure 1986, Henley *et al.* 1985). The $\delta^{2}H$ of basaltic magmatic water may vary from -60 to $-90‰$, while for water derived from andesitic magmas it may vary from -10 to $-30‰$. Depending on the geographic location of the geothermal water (either M1 or M2 as in Fig. 6.10), mixing between the geothermal reservoir water and magmatic water from basaltic magmas may result in the mixed water falling on the tie lines M1–B or M2–B. The case is similar with respect to mixing between reservoir water and magmatic water derived from andesitic magmas. The position of the $\delta^{18}O$ and $\delta^{2}H$ values of the mixed geothermal water on the tie line depends on the share of the magmatic water component. If the quantity of the magmatic water in the mixed water is more, then the plot will tend to move towards B or A.

6.4.4.4 *Steam separation*

Separation of steam influences the $\delta^{18}O$ and $\delta^{2}H$ values of geothermal waters. The effect is large when single stage steam separation occurs at near surface environments. In such situations a high rate of equilibrium between the steam and liquid is attained (Giggenbach 1971).

Let us consider the fractionation of oxygen and hydrogen isotopes between steam (s) and liquid (l):

$$\underset{R_0}{H_2{}^{18}O_s + H_2{}^{16}O_l} = \underset{R_1}{H_2{}^{16}O_s + H_2{}^{18}O_l} \tag{42}$$

Equilibrium constant α (or the fractionation factor) for this reaction is:

$$\alpha = R_1/R_s \tag{43}$$

Thus from equation (41) follows:

$$\alpha = 10^3 (R_1/R_{SMOW})/10^3 (R_s/R_{SMOW}) = 10^3 + \delta_1/10^3 + \delta_s \tag{44}$$

Since all isotope fractionation factors are close to unity:

$$\ln \alpha = \alpha - 1 \quad \text{for } \alpha \approx 1 \tag{45}$$

$$\ln \alpha \approx (10^3 + \delta_1/10^3 + \delta_s) - 1 = ((\delta_1 - \delta_s)/10^3 + \delta_s) \tag{46}$$

since:

$$(10^3 + \delta_s) = 10^3 \quad \text{for } \delta_s \ll 10^3 \tag{47}$$

it follows that:

$$10^3 \ln \alpha \approx \delta_1 - \delta_s \tag{48}$$

where δ represents $\delta^{18}O$ in the case of oxygen fractionation and δ^2H in the case of hydrogen fractionation.

At near surface environments, where single step steam separation occurs, the isotopic composition of the steam and liquid can be evaluated by:

$$\delta_{1+s} = \delta_s \, y_s + \delta_1 (1 - y_s) \tag{49}$$

where δ_{1+s} (also can be designated δ_o) is the original composition of the liquid before steam separation and y represents steam fraction in the original single phase containing both liquid and steam.

Using the enthalpies of liquid before steam separation (H_o), and that of liquid (H_1) and steam (H_s) after steam separation during the single stage, y_s can be calculated:

$$y_s = (H_o - H_1)/(H_s - H_1) \tag{50}$$

As described above, the isotopic composition of liquid and steam after steam separation can be calculated by using the following equations:

$$\delta_1 = \delta_o + y_s \, (10^3 \ln \alpha) \tag{51}$$

$$\delta_s = \delta_o - (1 - y_s) \, (10^3 \ln \alpha) \tag{52}$$

When the values of $\delta^{18}O$ and δ^2H of the liquid and separated steam are plotted on the $\delta^{18}O$ vs δ^2H diagram, the plots define a slope (S) governed by:

$$S = (\alpha_{2H} - 1)(\delta_o{}^2H + 1000)/(\alpha_{18O} - 1)(\delta_o{}^{18}O + 1000) \tag{53}$$

Therefore the $\delta^{18}O$ and the δ^2H values of the liquid after steam separation move to the right side of the MWL while the values in the steam that is separated move to the left of the MWL. When the steam separates below 220 °C, the $\delta^{18}O$ the δ^2H values of the residual liquid shows enrichment and above 220 °C, $\delta^{18}O$ in the liquid shows enrichment and δ^2H shows depletion (Fig. 6.11).

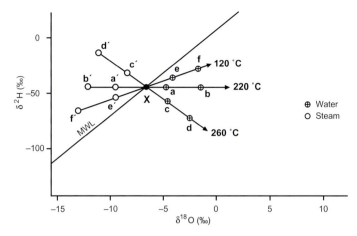

Figure 6.11.   Effect of steam separation on the oxygen-hydrogen isotopes in geothermal waters. 'X' represents geothermal water derived from precipitation (meteoric water). The geothermal water loosing steam at 120 °C tends to move towards positive values of $\delta^{18}O$ and δ^2H while the steam separated at this temperature becomes more negative with respect to $\delta^{18}O$ and δ^2H. Above 220 °C, the inversion of hydrogen fractionation factor occurs resulting in enrichment of $\delta^{18}O$ and depletion of δ^2H in steam separated geothermal fluids and enrichment of δ^2H and depletion of $\delta^{18}O$ in separated steam (modified after Truesdell *et al.* 1977, Giggenbach 1978).

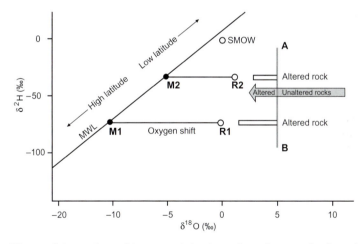

Figure 6.12.   Water-rock interaction and isotope variation in geothermal waters. Geothermal water (M1 or M2) reacting with wall rocks at temperatures above 220 °C gains only $\delta^{18}O$ and shifts horizontally (R1 or R2). Wall rocks loose $\delta^{18}O$ to the geothermal fluids and become depleted in $\delta^{18}O$. The degree of $\delta^{18}O$ enrichment in geothermal fluids and depletion in wall rocks depends on the temperature of the exchange reactions between the geothermal fluids and wall rocks (modified after Giggebach and Lyon 1977, Giggenbach 1978, Giggenbach and Stewart 1982, Truesdell and Hulston 1980, Henley *et al.* 1985).

δ^2H values of the liquid and the steam separated are constant at 220 °C and points fall on a line parallel to the $\delta^{18}O$ axis (Fig. 6.11), because fractionation for hydrogen does not take place ($\alpha = 1$).

   For example, let X represent the meteoric water recharging the geothermal reservoir in Figure 6.11. A small amount of steam separation at 120 °C results in the liquid shifting its composition to *e* while the separated steam shifts to *e'*. If the quantity of steam separated is greater at the same temperature, the liquid composition would shift to *f* while that of the steam shifts to *f'*. At 220 °C, due to the exchange of oxygen isotopes between the rock and water, steam separated geothermal water and steam do not show any change in δ^2H due to negligible content of δ^2H in the rocks.

Above 220 °C, inversion of the hydrogen fractionation factor occurs resulting in the enrichment of δ^2H in steam (Fig. 6.11).

6.4.4.5   *Interaction with reservoir or wall rocks*

All the silicate minerals in various rock types contain $\delta^{18}O$ between 5 and 10‰ (Faure 1986). A–B in Figure 6.12 defines the lower limit of $\delta^{18}O$ in the unaltered silicate minerals. Thus if any rock exchanges oxygen isotopes with the geothermal water, then the altered rock composition with respect to $\delta^{18}O$ shifts to the left of the line A–B indicating depletion of this isotope, while a corresponding increase in this isotope in the geothermal water will be observed. Thus, when geothermal water recharged by meteoric water of composition M1 reacts with rocks, the composition of the altered rock will shift towards the MWL while the composition of the geothermal water shifts to point R1. The magnitude of 'oxygen shift' depends on the temperature during the interaction and the amount of ^{18}O incorporated by the geothermal water. The percentage of depletion in the rock, in general, should be equal to the percent of enrichment in the geothermal water. The exchange reaction between water and rock takes places at or above 220 °C.

Similar to equation (48) above, the equilibrium constant α related to the rock (A) and water (B) can be expressed as:

$$10^3 \ln \alpha_{AB} \approx \delta_A - \delta_B \qquad (54)$$

As observed by Giggenbach (1988) from various geothermal fields like Wairakei, Ohaaki, Ngawha, Tonga and Salton Sea, the highest fractionation factor is shown by quartz while the feldspars show the lowest fractionation factor.

CHAPTER 7

Geophysical methods for geothermal resources exploration

> *"Of course, energy is not a single challenge that can be answered with a piece of technology as clean, appealing and profitable as an iPod. It is three intertwined challenges, each vexing and complex in its own right ..."*
>
> President S. Hockfield's opening remarks at MIT Energy Forum, May 2006.

7.1 GEOPHYSICAL TECHNIQUES

Before the drilling operation commences for exploitation of any geothermal resources, pre-feasibility investigations using geochemical and geophysical methods need to be carried out to have a better understanding of the subsurface geological and tectonic configuration of the region. In fact, geophysical exploration enables one to unravel the tectonic features and associated thermal regimes that host the geothermal reservoirs. Since drilling geothermal wells is expensive, geophysical methods bring down the cost to a certain degree. This is true in any underground exploration investigation for groundwater, mineral resources, mining activity, etc. Geophysical exploration methods help in indirectly obtaining the physical parameters (aquifer characteristics, depth of the reservoirs, temperatures of the reservoirs, area occupied by the geothermal reservoir, etc.) of the geothermal systems. Economical viability of geothermal resources can also be assessed through geophysical exploration.

Several geophysical methods are being applied to map tectonic, thermal, magnetic and gravity anomalies in the earth's crust that are commonly associated with geothermal systems. Thermal, electrical, magnetotelluric, magnetic, seismic, and gravity surveys are a few of the methods commonly employed for geothermal exploration. All these investigations are carried out on the surface over the geothermal regions.

7.1.1 Heat flow measurements

Geothermal provinces are characterized by having higher heat flow values than the average world value (63 mW/m^2). Heat flow values of subduction zones, deep continental rifts, active volcanic regions, and regions with rocks containing high concentration of radioactive minerals are anomalously higher than the global average and are loci of geothermal manifestations.

Surface heat flow values can be obtained from temperature gradient data and thermal conductivity of the rocks. Thermal conductivity can be measured in the laboratory on core or rock chip samples obtained from bore holes. Since rock contains a mixture of minerals, and because the thermal conductivity of each mineral varies depending on its chemical composition and structural state, it is difficult to obtain a single value of thermal conductivity. However, based on the conductivity of minerals it is possible to obtain the conductivity range of the rocks. For example, the thermal conductivity of magnesium rich olivine (Mg_2SiO_4) is around 4.65 W/m/K and the iron rich variety is around 3.85 W/m/K at 30 °C. Problems arise when iron minerals like magnetite or hematite veins traverse the rocks (Rybach *et al.* 1988, Clauser and Huenges 1995).

Laboratory determination of thermal conductivity values will be at variance from *in situ* conditions, since thermal conductivity is influenced by pressure, temperature, rock porosity and fluid composition.

Data on temperature gradients can be obtained from shallow bore wells drilled to a depth of about 1 m. Such measurements are less expensive and data on a regional scale can be obtained (Combs

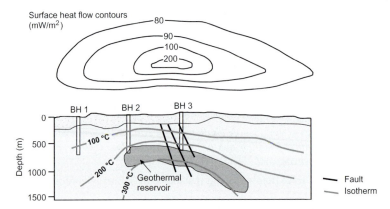

Figure 7.1.   Schematic cross section of a geothermal reservoir and its relationship with geothermal gradient and heat flow anomalies. Heat flow anomalies give good indication of the existence of a geothermal reservoir or a heat source that can be considered as an enhanced geothermal system (EGS).

and Muffler 1973). While taking measurements from deep bore wells, care should be exercised to avoid interference from topography, precipitation, and the groundwater table. Deep bore well data are generally available from agencies involved in groundwater exploration. The temperature measurements in bore holes can be obtained by thermocouple or thermistors. An alternate method of measuring the heat flow is by carrying out surface thermal surveys.

Heat flow values can be plotted from regional geothermal gradient data and conductivity values of the rocks. As shown in Figure 7.1, the geothermal gradient in bore wells BH2 and BH3 is much higher relative to bore well BH1. In the case of wet geothermal systems, the high geothermal gradient and high heat flow anomalies (Fig. 7.1) are associated with surface geothermal manifestation in the form of geothermal springs and fumaroles. In the case of hot dry rocks such surface geothermal expressions will be absent. The heat flow contours will certainly indicate the presence of a hot rock concealed > than 1 km below the ground. Surface heat flow values play a significant role in identifying hot dry rock provinces (Chandrasekharam and Chandrasekhar 2007).

7.1.2   *Electrical resistivity methods*

Electrical resistivity surveys are very effective tools in identifying wet geothermal systems since resistivity variations are directly related to the fluids and the host rocks (Bruno *et al.* 2000). The advantage of this method is that large areas can be surveyed with minimum cost. Direct current resistivity surveys with Wenner or Schlumberger electrode configuration are commonly used in geothermal exploration. Resistivity measurement of rocks is carried out with fixed electrode spacing. For exploring deeper depths, high current and large electrode separation are required, and this places a limitation on this method. However, Schlumberger electrode configuration, with electrode spacing of about 2 km to probe to a depth of about 1.5 km, is possible with little difficulty (Risk 1976).

In general the geothermal reservoirs are characterized by lower resistivity relative to the surrounding area. A resistivity map of a region gives indication about the area and depth of the underground geothermal reservoir. Resistivity depends on a number of factors like porosity of rocks and composition, and temperatures of the fluids. Resistivity shows an inverse relationship with temperature of the geothermal fluids and concentration of dissolved ions in the fluids. Thus, saline geothermal fluids register low resistivity as well as high-temperature reservoirs. Sometimes problems arise in regions with clays since the resistivity values more or less overlap the values shown by geothermal fluids. Such problems can be solved through field observations. Similarly, resistivity surveys over steam-dominated geothermal systems pose difficulty in interpreting the

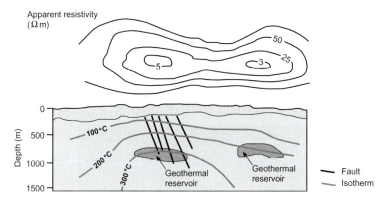

Figure 7.2.   Schematic cross section of a geothermal system showing surface resistivity anomalies. Low resistivity zones can be targeted for further exploration in a geothermal province.

anomalies. Fortunately this problem seldom poses problems because in steam-dominated fields a layer of steam condensate overlies the zone of steam, giving rise to low resistivity values. A schematic cross section of a geothermal system and the surface resistivity anomalies arising due to the geothermal reservoir are shown in Figure 7.2.

Depth limitations inherent in resistivity surveys can be offset by adopting the dipole-dipole method. This method is commonly used for locating deep-seated reservoirs containing geothermal fluids (Risk 1976, Bruno *et al.* 2000).

7.1.3   *Magnetotelluric survey*

The magnetotelluric (MT) method utilizes the earth's naturally occurring magnetic field and current. The magnetic field is measured over the surface of the earth. Basically this method measures the lateral and vertical electrical conductivity variation within the crust caused by the presence of ions or conductive solids (Simpson and Bahr 2005). Thermally excited solids (rocks) become highly conductive. Ions are present in the circulating geothermal fluids in the upper crust, or in the partial melts in the lower crust or in the mantle. This method can be applied in locating geothermal reservoirs within the crust. Basically this method images the earth's electrical resistivity structure at depths greater than 100 m. Low frequency method is used for locating large reservoirs located at greater depths, while audio-frequency is used for locating shallow reservoirs, and those on a regional scale. Since high heat flow values indicate the presence of subsurface geothermal reservoirs, heat flow variation in conjunction with MT surveys form an excellent tool in geothermal exploration. Based on the MT data, subsurface models can be developed using appropriate computer software. A schematic cross section of a typical geothermal system and the modeled MT survey results are shown in Figure 7.3. The low resistivity anomalous zones (Fig. 7.3a) lie above two geothermal reservoirs (Fig. 7.3b) with varying temperatures and different depths. Thus, MT surveys, in conjunction with field and other geophysical investigations described above, provide valuable information not only on the 3-D geometry of the geothermal reservoir, but also on the temperature variation of the reservoir with depth (Harinarayana *et al.* 2006).

MT surveys have been carried out extensively to identify the depth and extent of geothermal reservoirs, and also the temperature of the reservoirs before the commencement of the drilling program (Oskooi 2006, Harinarayana *et al.* 2006, Pastana de Lugao *et al.* 2002, Volpi *et al.* 2003).

The temperature-resistivity relationship indicates that resistivity decreases with an increase in temperature (Fig. 7.4) and this relationship is further influenced by the solute concentration of the fluids present in the reservoir (Nesbit 1993).

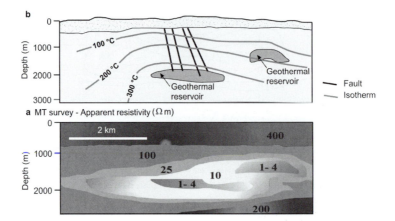

Figure 7.3.   (a) Schematic cross section of a geothermal system and model developed over the system using MT survey data. The resistivity values, in Ωm, of different subsurface strata are shown in (b) (modified after Harinarayana *et al.* 2006).

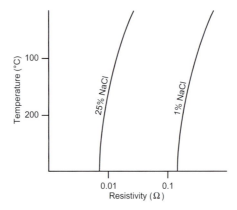

Figure 7.4.   Resistivity-temperature relationship for fluids of different ion-concentrations (modified after Nesbit 1993).

The reservoir temperatures can be estimated from resistivity values obtained from MT surveys using the following relationship:

$$T_z = T_s + (Q_s/k) - (A_s \times Z^2/2k) \qquad (1)$$

where T_z is the temperature (°C) at a depth Z (km), T_s the surface temperature (°C), Q_s the heat flow at the surface, k thermal conductivity of the material, and A_s the radioactive heat production (Harinarayana *et al.* 2006).

The temperature-resistivity relationship becomes very useful in locating hot dry rocks buried several kilometers below the surface. Mapping such hot dry rock regions will help in developing enhanced geothermal systems (EGS).

Other surface geophysical methods like refraction and reflection seismic, gravity, magnetic, and audio-magnetotellurics are also being used for geothermal exploration. These methods are site and location specific. They provide information on the sites to be identified for drilling exploratory bore holes.

7.1.4   *Geophysical well logging*

To understand the subsurface characteristics of the geothermal reservoirs, geophysical well logging methods are very important. Well logging provides information on the characteristics of the rocks penetrated by a bore well. Geophysical well logging is a method in which subsurface geological information is obtained from exploratory bore wells before the well is developed for producing geothermal fluids for power generation. In geophysical well logging, a probe is lowered into the exploratory bore well and necessary physical parameters of the geothermal reservoir are recorded for further interpretation (thickness, porosity, fracture pattern, pressure, and temperature of the geothermal aquifer, salinity and quality of steam, etc.). The logging tools contain logging sondes, and sensors (could be thermal, magnetic, electrical, radioactive and acoustic) that transmit data from deeper parts of the earth to a recorder located at the earth's surface. In this way a detailed evaluation of the respective properties along the bore hole axis can be carried out. These logging tools were initially developed by the oil industry for exploration of oil and gas, but with evolving technology, these tools are now being used extensively in geothermal exploration. A few selected bore hole logging methods, applicable for geothermal exploration, are discussed briefly in the following section.

7.1.4.1   *Gamma ray log*
The gamma ray log is useful mainly to evaluate the radioactive characteristics of the subsurface strata. This log, as the name implies, measures the intensity of natural gamma rays in the rocks traversed by the bore hole. When more than one exploratory bore hole is drilled, the data from the gamma ray logs can be correlated to delineate the area occupied by units of specific rocks that helps to estimate the volumetric capacity of the geothermal aquifer.

7.1.4.2   *Gamma-gamma density log*
The gamma-gamma density log is useful in interpreting the composition of the aquifer matrix, density of the geothermal fluids, and presence, or absence of gas phase in the geothermal reservoir. The principle behind this log is to emit medium energy gamma rays into the rocks. These gamma rays collide with the electrons in the minerals and back-scatter. The data thus obtained is useful in determining the density of the rocks and intensity of the fractures and estimate the steam quality in the reservoir.

7.1.4.3   *Acoustic log*
In an acoustic log, a transducer generates elastic waves that travel through the rocks. The time taken by the wave to travel a certain distance within the rocks is recorded. The data obtained through this log helps in interpreting the reservoir porosity, interconnected fractures, fluid channels, quality and type of fluids, pressure regime of the reservoir rocks, and elastic property of the rocks. This log is generally useful in the exploration of EGS.

7.1.4.4   *Neutron log*
The neutron log measures the concentration of hydrogen atoms present along the bore well axis. A radioactive source emits fast moving electrons that collide with the nuclei of the minerals present in the subsurface rocks. These neutrons, due to collision, slow down and are captured by elements such a chlorine, hydrogen and silicon that emit, as a result, high energy gamma rays. The emitted high energy gamma rays provide information on the hot rock horizons in the subsurface strata, quality of steam, and thickness of steam-bearing rocks.

7.1.4.5   *Temperature log*
A temperature log is widely used in geothermal exploration to measure the temperature gradient along the bore well axis. The data obtained from several exploratory drill holes can be synthesized to prepare subsurface isothermal contour maps that help to locate geothermal reservoirs with varying temperatures.

CHAPTER 8

Power generation techniques

"The opportunities for geothermal power to play a much larger role in overall energy production in the future require technical innovation, reduced startup costs, public education, and a level economic and regulatory playing field with other energy technologies."

Earthworm Tunneling Industry Encyclopedia, 2005.

8.1 OVERVIEW

Low-enthalpy geothermal resources can be used effectively for power generation using the binary cycle technique. This is also known as organic Rankine cycle (ORC). The turbo machinery component for this type of system is quite compact, easy to handle, and sometimes can be shifted easily. The basic principle behind this technique is to extract the heat from the geothermal fluids through a heat exchanger and transfer the heat to a low boiling point (BP) organic liquid (Fig. 8.1). The low boiling point liquid evaporates and develops enormous pressure sufficient to drive the turbine. The system is most suitable for low-enthalpy resources. Maximum heat from the geothermal fluids can be extracted through efficient heat exchangers and through selecting appropriate organic liquids.

Conventionally n-butane (BP: $-0.5\,°C$), isobutane (BP: $-11\,°C$), and toluene (BP: $110\,°C$) are used as organic fluids in the binary power system. N-butane or isobutane is most suitable for geothermal fluids below $200\,°C$, while toluene is suitable for fluids with $>200\,°C$ (Vijayaraghavan and Goswamy 2005). Compared to isobutane, n-butane has higher efficiency in ORC power plants (Fig. 8.2). Several other organic fluids (propane, pentane, hexane, heptane) have also been used as working fluids in ORC power plants.

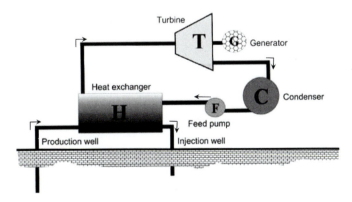

Figure 8.1.   Schematic diagram showing the principle of binary geothermal power plants and paths of the fluids.

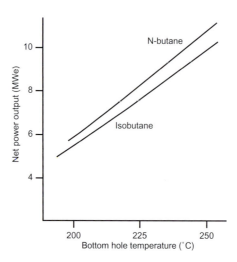

Figure 8.2.   Efficiency of power output by isobutane and n-butane in binary power plants (modified after Jacobs and Boehm 1980).

8.2   CRITERIA FOR THE SELECTION OF WORKING FLUID

The working fluid used in the ORC power plants should evaporate at atmospheric pressures and should have a low boiling point. These are two very important criteria that a working fluid should satisfy. General considerations for the selection of the working fluid are: it should be non-corrosive, non-flammable and should not react or dissociate at the pressures and temperatures at which it is used. Many organic fluids may not meet these requirements since most of them are inflammable and are not environmentally friendly. Working fluids in small binary power plants housed inside buildings may not necessarily meet all the specifications. Another advantage of using an organic working fluid is that when it enters the turbine low temperatures are obtained due to expansion of the vapors, and the turbine chamber needs not be under vacuum. After having said this, all the organic fluids fall under GHGs (greenhouse gases) domain and will attract the attention of environmentalists in the future. This problem has been overcome in recent years by using an ammonia-water mixture known as the Kalina cycle (see section 8.3). Since ammonia-water mixtures have a higher volatility, this aids in increasing the efficiency of the power plants (Vijayaraghavan and Goswamy 2005).

8.3   HEAT EXCHANGERS

Several types of heat exchangers like (1) direct contact, (2) plate, (3) coil tube, (4) panel, and (5) shell and tube heat exchangers are available for use in binary power plants. The choice of the heat exchanger for a binary power plant depends on several factors; a few of these are listed below (Dart and Whitebeck 1980):

- Composition of the fluid, flow rate, inlet and outlet pressure, and temperature.
- Working fluid property, e.g. its boiling point.
- Fouling characteristics of the fluid.
- Working and geothermal fluids flow path.
- Heat transfer data.
- Surface area of contact of the fluid.

Shell and tube exchangers are best suited for extracting the maximum amount of heat from geothermal fluids and are being widely used. A detailed account of the material design of heat exchangers is given by Dart and Whitebeck (1980).

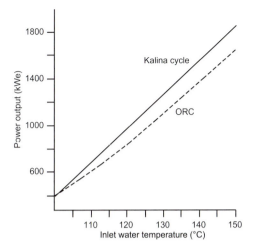

Figure 8.3. Comparative analysis of power generating capacity by Kalina cycle and ORC. Above 100 °C Kaliana cycle has greater power generating capacity (modified after Valdimarsson and Eliasson 2003).

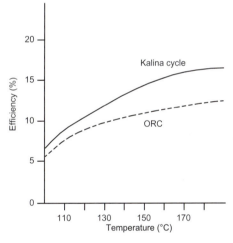

Figure 8.4. Comparative analysis of efficiency of Kalina cycle and ORC. Kalina cycle is more efficient in generating power from binary power plants compared to ORC. For low-enthalpy geothermal systems the Kalina cycle is more suitable than ORC (modified after Valdimarsson and Eliasson 2003).

8.4 KALINA CYCLE

This cycle, named after Dr Alexander Kalina who invented it (Kalina 1983), utilized a simple mixture of ammonia and water that was used to run engines. This cycle is also known as the ammonia-water cycle. Later, the design was modified to suit several industrial applications. Waste heat from industries (for example, cement industry) can be utilized to generate power using the Kalina cycle. Ammonia and water have similar molecular weights (ammonia: 17 kg/kmol; water: 18 kg/kmol), are soluble in each other and can be separated with ease. Both these liquids have different boiling temperature (ammonia: −33 °C; water: 100 °C) and thus, the mixture, depending on the mixture ratio, evaporates over a wide range of boiling temperatures and is best suited to generate power from low-enthalpy geothermal water by making slight alteration to the steam turbines (Kalina and Leibowitz 1987). Ammonia is not expensive and is used commonly in

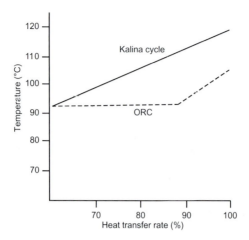

Figure 8.5.   Comparative analysis of heat transfer by Kalina cycle and ORC (modified after Kalina 1984).

absorption refrigeration. Ammonia does not deplete ozone, and thus does not contribute to global warming.

The economic considerations were studied in details by several workers (Kalina *et al.* 1991, Kalina and Leibowitz 1989) and the cycle's applicability to geothermal power plants evaluated (Kalina and Leibowitz 1989, Lazzeri 1997). Although the initial investment in the Kalina cycle is higher compared to ORC, this higher cost is offset by its ability to generate more power (Fig. 8.3) efficiently (Fig. 8.4). Though the Kalina cycle requires a larger heat exchanger, the heat transfer is far greater than that in the ORC (Fig. 8.5). However, the turbine size is small in Kalina and less expensive compared to ORC (Jonsson 2003). The advantage of the Kalina cycle over ORC is its efficiency in generating 20–40% more power from the same fluid temperature due to 100% heat transfer from the geothermal fluid to the secondary fluid. Payback is estimated at 1.5 years *versus* 6.5 for Rankine cycle systems used as bottoming cycles, and based on $0.06 per kWh electricity cost (Kalina 1984).

CHAPTER 9

Economics of power plants using low-enthalpy resources

"Today, over two billion people in developing countries live without any electricity. They lead lives of misery, walking miles every day for water and firewood, just to survive. What if there was an existing, viable technology, that when developed to its highest potential could increase everyone's standard of living, cut fossil fuel demand and the resultant pollution...."

P. Meisen, President, Global Energy Network Institute (GENI), 1997.

9.1 DRILLING FOR LOW-ENTHALPY GEOTHERMAL RESERVOIRS

The selection of drilling equipment depends on the characteristics of the geothermal reservoirs. Drilling into soft sediments is much easier than drilling into hard rocks, and the drilling equipment used for oil wells is well suited for such rocks. The difference between oil wells and geothermal wells is that in the case of geothermal wells, in addition to pressure (like in oil wells), temperature criteria and the corrosive nature of the fluids, need to be considered. Hence, drilling geothermal wells is a specialized technique and demands specialized expertise and skills. Rotary drilling equipment, generally employed for oil wells, is used for geothermal wells. For shallow wells the drilling rig can be mounted on a truck and transported with ease. Such drilling equipment is well suited for transportation to remote places in the development of small power plants. An additional advantage is that it can be air lifted to rural sites easily.

Unlike oil or gas wells, drilling geothermal wells poses certain inherent problems. Weight on the drilling bits in geothermal wells is required to be low to increase the bearing life of the bit. This low weight reduces the penetration rate. In addition to this, depth and elevated environment temperature constrain the drilling rate. Three or less wells will need to be drilled for small power projects, as described earlier. The drilling site can be prepared accordingly, and all three wells may be made from a single drill pad (DiPippo 2005).

Down the hole hammer (DTH) method sometimes becomes convenient for drilling shallow wells, especially in hard rocks (Sanner and Anderson 2002). If the rocks are unstable, then a rotary drill becomes the best option. Drilling fluids play an important role in the drilling operation, by increasing the life of the drill bit, lubricating the drill string, and stabilizing the bore hole.

At present there is a variety of drilling fluids available. Bentonite, cellulose, polyacrylamide, barite, and foam generators are a few drilling fluids used depending on the rock type and site specification (Sanner and Anderson 2002). In geothermal drilling the drilling fluids get heated and need to be cooled before they are re-circulated.

9.2 DRILLING COST

The most important components that dictate the cost of drilling are: (1) cost while drilling is in progress, (2) cost while drilling unit is idling due to repairs, maintenance, well logging, etc., (3) cost associated with the completion of the well, (4) other costs associated with shifting of the drilling unit, material transport needed for drilling, and miscellaneous expenses, (5) depth of bore well, (6) nature of the rock being drilled and the number of casing strings required, and (7) the diameter of the well (DiPippo 1980, Hance 2005). Further factors that control the drilling cost are unforeseen conditions that may arise during the drilling process (e.g., well blow outs). During the evaluation phase of exploration, the "wildcat" wells have a success rate of about 25% (Hance 2005). During this phase, the cost is about 77 US$/kW (Hance 2005). For small geothermal systems, this cost will be much lower since well sites are chosen based on prior information from the site on an existing data base. According to various developers, drilling a geothermal well takes from 25 to 90 days. Deeper wells may take a much longer time. This drilling time is only an indicator when determining the budget needed for drilling.

When the subsurface permeable geological rock pressure is lower than the hydrostatic pressure, or overburden pressure, this may cause a loss of drilling fluids in the productive zone. Drill bits entering such highly permeable or highly fractured zones may cause a loss of drilling fluids. The well may demand additional casing and, or a change of drilling fluids. Whether to drill further to greater depths or not is a major decision to be made by the developer. Penetration to greater depths may permanently plug the permeable shallow resource zone. Such factors apparently add to the cost of drilling. As mentioned above, if the geothermal fluids are highly corrosive, the casing should use corrosion resistant material and cement. A titanium liner may protect the casing from corrosion, thus increasing the cost of the well. The estimated cost of titanium pipes of ~13 inch diameter is around US$ 3500/m (Hance 2005). If the reservoir contains non-corrosive geothermal fluids, this problem may not arise. Geothermal fluids associated with non-volcanic zones (see Chapter 5) may pose fewer problems compared to those associated with active volcanic zones.

9.3   DRILLING COSTS *VERSUS* DEPTH

M/s GeothermEx Inc. in a consultation report submitted to the California Energy Comission (GeothermEx 2004), conducted a detailed analysis of drilling cost *vs* well depth relationship. The analysis is based on a strong data base of 182 geothermal wells (free flowing as well as wells fitted with downhole pumps) from 17 geothermal fields, drilled between 1985 and 2000 in North America (Geysers and Salton Sea), Central America (El Salvador, Guatemala), and the Azores. The drilling cost was normalized to 2003 US$ values taking into account the inflation rate using the producer price index (PPI) for onshore oil and gas drilling. Production as well as injection well cost, pad construction cost, cost of mobilization, and de-mobilization of the drilling equipment were included in this analysis. The diameter of the wells within the reservoir is in the range of 8½ to 12¼ inches (21 to 31 cm). The analysis shown in Figure 9.1 gives comprehensive insight

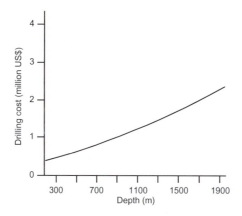

Figure 9.1.   Relationship between drilling costs and depth of geothermal wells (modified after GeothermEx 2004).

into the drilling cost *vs* depth relationship of geothermal wells. The curve shown in Figure 9.1 is the best fit line drawn based on a second order polynomial with a correlation coefficient of ~0.56 indicating that 56% of the variance in drilling is related to depth. Since the depth range in the above analysis varies from 200 to about 3500 m with resource temperatures ranging from 110 to 236 °C, this graph can be used as an indicator (it may be considered as an upper limit) for estimating the costs of drilling for low-enthalpy geothermal wells.

In general, drilling costs vary with operation time of the rig (25 to 90 days). Shallow wells (~1 km) may cost about 2 million US$ while deeper wells may cost up to 5 million US$. If the shallow wells are located in hard rock then the cost may exceed US$ 2 million.

9.4 WELL PRODUCTIVITY *VERSUS* RESERVOIR TEMPERATURE

Reservoir temperature and pressure influence the flow rate, and this in turn determines the productivity of the well and hence the cost of the exploitation. A high flow rate may reduce the number of wells while a lower flow rate may demand more wells to supply sufficient fluids to run the power plant. In a similar analysis described above, GeothermEx (2004) brought out a relationship between the reservoir temperature of a geothermal field and the productivity of a well. The data base is the same as that described above. The productivity per well, estimated by dividing the plant capacity by the number of active production wells, was correlated with the average temperature of the aquifer (Fig. 9.2). The power plant capacity used in developing this relationship varies from 0.7 MW$_e$ to 300 MW$_e$. The gross capacity per well, with depths varying from 200 to 3500 m, range from 0.7 to 8.9 MW$_e$. The temperature of the reservoir considered here varies from 110 to 236 °C. Quite often, wells tapping low-enthalpy reservoirs need downhole pumps to feed the binary power plants. Downhole pumps may control the pressure in the reservoir, thus preventing emergence of vapor phase; however the productivity of such wells is controlled by the capacity of the pump.

9.5 POWER PRODUCTION *VERSUS* WELL HEAD TEMPERATURE AND FLOW RATE

Bloomster and Maeder (1980), while describing the economic considerations of power plants and electricity production for high-enthalpy geothermal systems, brought out a relationship between the unit cost of electricity and well head temperature (Fig. 9.3a) and flow rate (Fig. 9.3b) in a well. Their analysis shows that when the well head temperature increases from 150 to 250 °C, the power cost is reduced to one third. This power cost included the cost of the power plant as well.

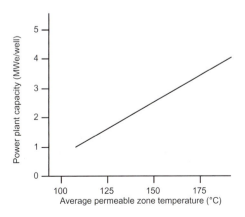

Figure 9.2.   Geothermal power plant capacity *vs* reservoir temperature (modified after GeothermEx 2004).

Thus for a well flowing at a constant flow rate, the unit cost of power is controlled by the well head temperature. The power cost for low-enthalpy systems apparently will be lower than these values since the cost of drilling shallow wells will be lower by a factor >2 (see section 9.3).

In binary power plants using low-enthalpy fluids, power production depends on several factors discussed above. The flow rate and resource temperature are the main deciding factors in estimating the productivity of the well. Recent analysis on the Raft river geothermal field (Sanyal *et al.* 2005), clearly demonstrates that proper selection of submersible pumps in conjunction with advances made in binary technology will aid in increasing the efficiency and productivity index of small power plants with resource temperatures varying from 140 to 146 °C. Important factors that provide certain guidelines in planning small power projects from low-enthalpy resources are discussed in the following sections and in Chapter 10.

9.5.1   *Raft river geothermal field*

The Raft river valley, Idaho, is a late Cenozoic rift valley filled with Tertiary sediments and volcanics overlying Paleozoic and Precambrian rocks. The Paleozoic rocks include quartzites, and metamorphic rocks like schists. The Precambrian rocks are mainly quartz monzonites (Applegate and Moens 1980). Gravity, magnetic, refraction seismic, resistivity and audio-magnetotellurics, selfpotential, and telluric surveys indicate the thickness of the Cenozoic sediments to be 2000 m (Applegate and Moens 1980). Bore wells drilled to depths varying from 1375 to 1520 m yielded fluids with a flow rate of about 41 to 80 l/s. The temperature measured at the bottom of wells varies from 133 to 147 °C. It is apparent that the Cenozoic and Paleozoic rocks are the main geothermal reservoirs in this valley. The reservoir temperatures estimated based on chemical geothermometers (water and gases) vary from 136 to 142 °C (McKenzie and Truesdell 1977, Nathenson *et al.* 1982), indicating chemical equilibrium conditions in the reservoir between the fluids and the rocks.

Due to the non availability of advanced binary technology and due to limitations in the efficiency of submersible pumps, at that time, this field was not developed for power generation. But now with the advances made in submersible pumps and binary power generation technology over the past two decades, this field is now prepared to generate >17 MW$_e$. Well productivity and issues related to pumping were analyzed based on the transmissivity of the reservoir, storage capacity of the reservoir, interference from the wells, inlet and re-injection temperature of the fluids, and specific heat of the fluids. These parameters were used to calculate the power capacity of this field.

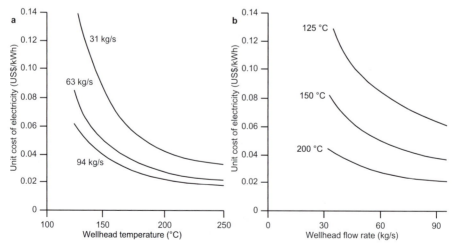

Figure 9.3.   Relationship between unit cost of power and well head temperature (a) and flow rate (b) in geothermal wells (modified after Bloomster and Maeder 1980).

The fluid required to generate 1 kW$_e$ was estimated using the utilization efficiency and maximum thermodynamically available work per unit mass of fluid. The calculation shows that for generating 1 MW$_e$ of power from a fluid of temperature 140 °C, the required flow rate is about 27 l/s while for fluids with 146 °C, the rate of flow should be maintained at about 25 l/s. The flow rate estimated is lower than that given in the Figures 9.3a, b. The Raft river geothermal project clearly has demonstrated that all the low-enthalpy geothermal fields are potential targets for generating power, especially in developing countries.

9.6   HIGH-ENTHALPY *VERSUS* LOW-ENTHALPY POWER PLANTS

In the case of high-enthalpy geothermal power plants, the investment costs include surface equipment costs (power plant, steam gathering system, etc.) and subsurface investment costs (drilling cost, number of wells drilled, etc.). For larger power plants, (>5 MW$_e$) the surface cost is a small part of the total cost of the project. The major cost involved in lager plants is the subsurface cost. For example, according to the Namafjall power project cost analysis (Stefensson 2002), 37% of the total cost of the project is taken by the subsurface cost of a 20 MW$_e$ plant. For large power plants, the number of wells required to be drilled will be greater to maintain the required flow rate to the plant, increasing the surface cost. The surface cost analysis on two high-temperature fields in Iceland (Bjarnarflag and Krafla) indicates that the surface cost and size of the power plant define a well-fit linear relationship (with correlation coefficient $R^2 = 0.97$) (Stefansson 2002). This relationship is defined by the equation (Stefansson 2002):

$$\text{Surface cost (million US\$)} = (-0.9 \pm 4.6) + (1.0 \pm 0.1) \times \text{MW}$$

This equation is valid for Iceland and need not be applicable to all fields. The geographic location of the geothermal fields also contributes significantly to the surface costs. However, this equation can be a guide in estimating surface costs for geothermal plants of 20–60 MW$_e$ capacity. Besides the production wells, large power projects need re-injection wells to inject the utilized geothermal fluids back into the aquifer. Hence, the power project development cost includes the production

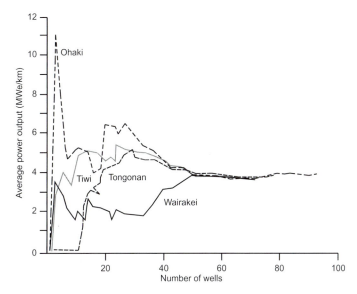

Figure 9.4.   Average power output per drilled kilometer of well in high-enthalpy geothermal fields (modified after Stefansson 2002).

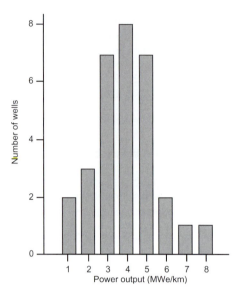

Figure 9.5.   Average output of power from thirty-one wells from high-enthalpy geothermal provinces of the world (modified after Stefansson 2002).

Table 9.1.   Average values for 31 high-enthalpy geothermal fields.

Average MW$_e$ per well	4.2 ± 2.2
Average MW$_e$ per drilled km	3.4 ± 1.4
Average number of wells before maximum yield achieved	9.3 ± 6.1

Source: Stefansson 2002.

and injection well costs. In the case of small power plants, re-injection is not an important part of the project development.

However, subsurface cost estimates for 31 high-temperature geothermal fields located in a wide range of geographical locations appear to be similar after the initial exploratory phase (Stefansson 1992). The rate of flow in exploratory/production wells may decrease or increase over a period of time and provisions should be made to maintain the electric power production rate by drilling additional wells in case the flow rate decreases. This decrease in the flow rate may be due to the wells drawing fluids from the same aquifer. The flow rate will stabilize over a period of time, thus stabilizing the production capacity below the initial value. Therefore, the cost of power production will stabilize over a period of time. This is true in all the high-enthalpy geothermal provinces of the world, as shown in Figure 9.4. The average output as shown in Figure 9.4 is about 4 to 5 MW for each drilled km. This is only an indicator, but to achieve actual output (4 to 5 MW$_e$) the number of wells required may vary. The final cost estimate for installed MW$_e$ power should be made after the well stabilizes and produces a fairly uniform flow rate to the generator. An amount of US$ 37 million for a 20 MW$_e$ plant or US$ 1750 per installed kW was reported by Stefansson (2002) for a power plant in Iceland.

The yield of the well per drilled kilometer, from the 31 high-enthalpy geothermal provinces, is presented by Stefansson (2002) on a histogram shown in Figure 9.5 and the actual values as reported by Stefansson (2002) are given in Table 9.1.

Using the criteria shown in Table 9.1, and assuming the average depth of a well to be 1.5 km, drilled at a cost of 1.5 million US$, the cost estimated for 1 MW$_e$ power project is about 0.29 million US$/MW$_e$. The estimated cost, including both surface and subsurface costs for a 40 MW plant, varies between 45 and 79 million US$. This works out to be 1225 to 1975 US$/kW.

CHAPTER 10

Small low-enthalpy geothermal projects for rural electrification

"All developing countries have a stake in energy sustainability, but in the Asian region especially, large populations aspiring to greater prosperity will strongly test our ability to deliver energy sustainability. Collectively, we will need to distill all available wisdom on the policies, market structures, pricing arrangements and technologies that can lead us to our goals. These issues are also the ones, which pre-occupy industrialized countries."

Welcome Address, secretary general, World Energy Congress 2004.

10.1 DEFINITION OF SMALL GEOTHERMAL POWER PLANTS

Small geothermal power plants have been defined as those that can generate less than 5 MW$_e$ (Lund and Boyd 1999, Vimmerstedt 1999). The advantage of small power plants is that they can utilize low-enthalpy geothermal fluids in conjunction with Kalina cycle or ORC binary power generation units. It is estimated that 1 MW$_e$ plant could serve about 20,000 households (Cabraal *et al.* 1996) assuming that the demand for electricity per person at off-grid sites will be of the order of 0.5 kW. In the past it was difficult to get financial support for such power plants due to the high cost compared to large plants (Lund and Boyd 1999). But due to inherent problems that large urban areas are facing with respect to land, water, and power, much of the future infrastructural development in developing countries will be in rural areas, creating huge economic development. It is then unlikely that such small power plants would face financial problems in the next decades. A list of power plants generating <5 MW$_e$ in different countries is shown in Table 10.1.

Most of the existing small plants were installed as pilot plants to prove the generating capacity of the geothermal province or to initiate the process of installing larger plants (e.g., Dieng, Fang; Ramingwong and Lertsrimongkol 1995). These small power plants are at present associated with larger plants shown in the Table 10.1 except those that are installed in remote rural areas in China (Tibet), Neuquén (Argentina) and Chena (Alaska). In case the installation of larger plants is delayed, these small power plants remain as small units serving the rural population.

Table 10.1.   Small geothermal plants currently in operation in various countries (modified after Vimmerstedt 2002). RT: Reservoir temperature; Expan. exp.: Expansion expected.

Power plant name	RT °C	Country	Site	Geothermal field	MW	Status
Amedee Geo.	104	USA	California	Amedee HS	2	Operating
Fang GT demo plant	116	Thailand	Fang	Fang	<1	Expan. exp.
Bouillante	160	France	Guadeloupe		4	Operating
Copahue power stat.	166	Argentina	Neuquén prov.	Copahue	1	Not operating
Empire geother. project	137	USA	Nevada	San Emidio KGRA	5	Expan. exp.
Nagqu	110	China	Tibet		1	Operating
Pico Vermelho	200	Portugal	Pico Vermelho		3	Expan. exp.
Kirishima hotel	127	Japan	S Kyushu		<1	Near larger site
Wabuska	104	USA	Nevada		2	Operating

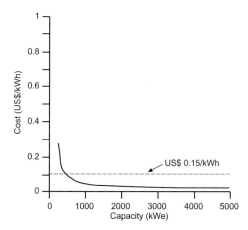

Figure 10.1.   Relationship between power capacity and cost of electricity. The cost curve was drawn based on the cost estimated for 2 wells drilled to a depth of 1.6 km with a plant capacity factor of 0.8 without overhead expenses like taxes or debt (modified after Vimmerstedt 2002).

10.2   CHARACTERIZATION OF RESOURCES AND COST REDUCTION

The exploration methods are similar, but vary in magnitude, in the case of low-enthalpy and high-enthalpy geothermal systems. Geochemical (Chapter 6) and geophysical (Chapter 7) methods play an important role in characterizing the resources. The main cost involved in geothermal exploration is drilling. Large projects (related to high-enthalpy resources) need several exploratory drill holes that may or may not be developed into production wells. This cost can be reduced by drilling more shallow drill holes (exploratory cum production) in the case of low-enthalpy resources. The number of drill holes to be drilled depends on the reservoir's capacity to generate power. The cost of drilling reduces with increase in plant production capacity as shown in Figure 10.1. Therefore, small power plants using low-enthalpy geothermal resources are quite cost effective and the electricity cost will be affordable to the rural population without government subsidies.

The current drilling technology for small power plants is focusing on popularizing slim holes. The slim holes are small diameter wells (15 cm diameter) compared to conventional wells (21 cm diameter). Slim holes are more cost effective if the well is under artesian conditions. Downhole pumps can be used under non-artesian conditions but the economy depends on the electricity produced and the capacity factor.

Economically, it is advantageous for developing countries to utilize locally available expertise in resource characterization. It is easier for these countries to mobilize funds for small power projects and avoid problems associated with data sharing and intellectual property rights. When experts are invited from outside the country, laws should be clearly drawn at the beginning of the project about data sharing and "intellectual property right" (IPR) issues for the smooth implementation of the project. As far as possible, for small-scale power development projects, existing data from wells drilled for oil, water and gas may be used to reduce the cost of exploration.

10.3   ENERGY NEED FOR RURAL SECTOR

Before initiating small power plants for rural areas, it is necessary to assess the present energy status and future energy demand. It depends on the income of individual households. In rural areas the households are classified as low, middle, and high income groups depending on their earnings. Table 10.2 (World Bank 1996) gives an idea about the energy use and future requirements.

Table 10.2.   Rural energy pattern in developing countries (modified after World Bank 1996).

End use	Household income		
	Low	Medium	High
Cooking	Wood, dung	Wood, dung, kerosene, biogas	Kerosene, biogas, LPG, and coal
Lighting	Kerosene	Kerosene, and gasoline	Kerosene, diesel, solar, and gasoline
Space heating	Dung (often none)	Wood, bio-residues, and dung	Wood, residues, dung, and coal

Table 10.3.   Cost of power generated from diesel generators in rural areas.

	US$
Generator (2 kW)	1300/kW
Fuel cost	0.1352/kWh
O & M	582/year

(Source: Vimmerstedt 1998).

As shown in Table 10.2 diesel, kerosene and gasoline are the main sources used for electricity generation. The current cost of electricity generated from diesel is given in Table 10.3.

Assuming a capacity factor of 40% and 20% fixed charges, the average unit cost of electricity for a system size from 2 to 10 kW in a large section of Caribbean and Latin American countries varies from 0.07 to 0.40 US$/kWh with an average cost of 0.25 US$/kWh (Vimmerstedt 1998). In fact all the Latin American and Caribbean countries have sizable low-enthalpy geothermal resources lying unutilized (Chandrasekharam and Bundschuh 2002). Diesel power cost is driven by the fuel price and transportation cost. In some countries, like India, diesel is transported to rural areas by airplanes thus increasing the power generation cost by several-fold. For example, in Leh, Ladakh, the unit cost of diesel generated electricity is 0.16 US$/kWh (Chandrasekharam *et al.* 2004). Hence, even if the diesel generators perform as per their manufacturing design, the unit cost of electricity cannot be controlled over a long period time. Additional costs are related to the O & M of the diesel generators to give maximum performance.

There are other extreme cases, like certain rural villages in Ladakh, India, where the rural population has not seen an electric bulb, even after 60 years of independence. Such drastic neglect of rural communities severely affects the socio-economic growth of the children and future generations of the under-privileged (Chandrasekharam *et al.* 2004). Under rural electrification schemes, solar powered cells were provided to supply electricity to individual homes. The higher income households may have the financial capacity to maintain such systems but middle and low income households that encompass a large percentage of the community, may not have the financial capacity to maintain such systems. Implementing agencies may have one time grants for rural electrification through photovoltaic schemes, but these agencies are not geared to render services after a couple of years (Chandrasekharam *et al.* 2004). The average age of the photovoltaic systems is about 2.4 years and failures outnumber successes due to the inhospitable terrain, harsh winters, and lack of infrastructure for maintaining the system (Chandrasekharam *et al.* 2004). Further, as described in section 10.4, the unit cost of electricity from photovoltaics is much higher compared to diesel if the subsidies by the government are removed.

As shown in Table 10.4, more than two billion people in the world have no access to modern technology and have no advanced energy sources like nuclear, geothermal, solar, or wind (Barnes *et al.* 1997). These communities still use traditional energy sources, like those listed in Table 10.2. These sources are not cost effective, inefficient, and cause severe respiratory diseases. In some countries, like India, even if the source is available, the local government does not take the initiative to exploit this source for the benefit of the rural community. A typical example is that of Leh, that

Table 10.4.   Urban and rural people connected to electricity in developing countries.

Region	Urban % 1970	Urban % 1990	Rural % 1970	Rural % 1990
North Africa and Middle East	65	81	14	35
Latin America and Caribbean	67	82	15	40
Sub-Saharan Africa	28	38	4	8
South Asia	39	53	12	25
East Asia and Pacific	51	82	25	45
All developing countries	52	76	18	33
Total served (millions)	320	1100	340	820

Source: Barnes *et al.* 1997.

Table 10.5.   Unit cost of electricity generated from low-enthalpy based small power plants (DePippo 1999).

Net power (kW)	Capital cost US$/kW Resource temperature (°C) 100	Capital cost US$/kW Resource temperature (°C) 120	Capital cost US$/kW Resource temperature (°C) 140	O & M cost US$/year
100	2786	2429	2215	21010
200	2572	2242	2044	27115
500	2357	2055	1874	33446
1000	2143	1868	1704	48400

is located about 150 km north of the Puga geothermal field (see Chapter 5). As described above, the local government spends huge amounts of money to provide electric power using diesel that is transported from far off places. The transportation cost of the fuel is added to the cost of electric power and the consumer has to pay the transportation cost. Generation of electric power from fossil fuels not only poses environmental problems, but also restricts the low and middle income households access to electric power.

During the last several decades rural populations did not see much progress with respect to electricity connectivity. The worst affected communities are those from South-East Asia, Sub-Saharan Africa, Latin America and the Caribbean (Table 10.4). It is estimated that more than 65% of people from developing countries live without electricity. Demand for electricity will increase by several factors, especially in developing countries (see sections 2.1 and 2.2 in Chapter 2). With the advancements made in drilling and power plant technologies, these countries today are in a position to exploit available low-enthalpy geothermal resources to enhance their socio-economic status. Low-enthalpy geothermal resources have not been considered for generating electric power during the last several decades. But now, binary technology using the Kalina cycle or ORC can utilize geothermal resources with temperatures as low as 74 °C.

Compared to the cost of electricity produced from photovoltaic units and diesel generators, the cost of electricity generated from small geothermal power plants using low-enthalpy sources is cost effective (Table 10.5) and has a positive effect on pristine environments by reducing the CO_2 levels in the atmosphere.

The unit cost of electricity from small power plants is much lower than the 0.25 US$/kWh supplied through diesel generators. As described above, the unit cost of electricity generated from small geothermal power plants ranges from 0.04 to 0.08 US$/kWh (see section 10.7). For example, the 300 kW binary geothermal power plant in Fang (116 °C) supplies power at the rate of 0.063 to 0.086 US$/kWh (Lund and Boyd 1999) while the unit cost of electricity generated from diesel is about 0.22 to 0.25 US$/kWh. Similarly, the Tu Chang binary power plant in Taiwan sells geothermal power at the rate of 0.04 US$/kWh to the Taiwan Power Company.

The above examples amply confirm the fact that small geothermal power plants are very cost effective for rural areas. When such power plants are linked to direct use applications, like greenhouse cultivation, dehydration, and aquaculture, the unit cost of electricity can further be lowered.

Thus, low-enthalpy geothermal resources have a greater role to play in rural areas of developing countries. The merit of such low-enthalpy power relative to the high-enthalpy systems can be evaluated based on the currently existing technologies in such remote places.

10.4   MARKETS FOR SMALL POWER PLANTS

Globalization, growing rural market demands, and increasing literacy among the rural youth are all transforming the developing countries' growth profile. With these changes the demand for electricity is growing rapidly. Access to better energy technology could aid the growth profile of rural communities. Small geothermal plants could be one of these technologies. Rural electricity services can be improved by installing individual systems, national grids, and mini-grids. In the case of remote areas where supplying power is not economical due to transmission losses and length of transmission line costs, small-scale geothermal power projects become handy. Moreover, it must be considered that in developing countries rural people have a low per-capita electricity demand relative to the urban population. This market might best be served by many small power plants rather than a few larger ones. According to estimates made in many areas of Latin America, the Caribbean, the Philippines, and other developing regions, and assuming an electricity requirement of 50 W (for lighting) per household, a 1 MW plant can serve about 20,000 households (Vimmerstedt 1998). The drilling cost, as shown in Figure 10.1, may increase the unit cost of electricity. The best economically viable solution is to establish mini-grids to supply electricity to a large section of the population by generating 2–3 MW_e.

Electric sector reforms are transforming the potential owners and operators of small geothermal projects from public utilities to private power producers. These reforms are intended to improve the overall economic efficiency of the electric sector and may open new opportunities for small geothermal projects in this more competitive market. Systems for use of geothermal energy have proven to be extremely reliable and flexible. Binary power plants are online an average of 97% of the time. Geothermal plants are modular, and can be installed in increments as needed. Because they are modular, they can be transported conveniently to any site. Both baseline and peaking power can be generated. Construction time can be as little as 6 months for plants with capacities in the range of 0.5 to 10 MW_e and as little as 2 years for clusters of plants with total capacity of 250 MW_e or more (Chandrasekharam 2001a).

Though drilling cost defines the production capacity of power plants (see Fig. 10.1), small power plants with a generating capacity below 1 MW_e are successfully operating in several areas along the western part of the USA and other regions of the world where low-enthalpy power plants are installed. As mentioned above, because of the shallow depth of the wells the drilling costs reduce drastically thereby making such plants more economical. For example, Wineagle Developers geothermal site in California, USA, is generating 750 kW_e from two modular binary plants. These two binary units are operating with a gross efficiency of 8.5% and with a capacity factor of 109% (Lund and Boyd 1999). Similarly, Amedee geothermal venture binary plant in California was commissioned in 1988. This plant generates 1.5 MW_e from two binary units with a fluid temperature of 104 °C. The geothermal water, drawn from a well of depth of 260 m, is supplied to the plant. The fluid flow rate is about 205 l/s (Lund and Boyd 1999). This plant uses R-114 as the working fluid and is remotely monitored by a telephone line. Such plants are best suited for rural areas in developing countries. Under the present technology, such plants can operate using the Kalina cycle, thus reducing the cost of unit power and the total project cost as well. The flow rate of the fluids is another factor that decides the cost of the power plant (see section 10.5).

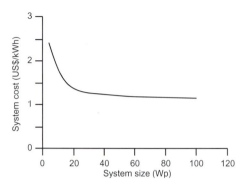

Figure 10.2.   Unit cost of electricity generated through a photovoltaic system assuming levelized capital for a capacity factor of 0.2 and fixed cost rate of 0.20 (modified after Vimmerstedt 2002).

Other power systems that generally compete with geothermal systems include solar photovoltaic, diesel, hydropower, oil, and gas turbines.

Solar photovoltaic cannot compete with geothermal since the unit cost of electricity is much higher than that of geothermal small power plants (Fig. 10.2). Photovoltaic based power systems may not be able to supply low cost electric power for the next few decades (World Bank 1999).

Diesel generators, as described in section 10.3, are extensively used for rural electrification purposes. In certain localities, for example in Leh town (Ladakh province, India), diesel is transported from long distances and the cost of transportation is embedded in the unit cost of electricity that will have to be borne by the consumer. Furthermore, diesel generates huge amounts of greenhouse gases that pollute the pristine climate of the region. The cost of electricity generated from diesel in Leh town, with government subsidy, is higher by a factor of 4 compared to the unit cost of electricity supplied in other parts of the country. If the government subsidy is removed, then the unit cost of electricity goes beyond the paying capacity of the rural population.

Hydroelectric power projects attract environmental issues and land submergence, thus displacing considerable numbers of homes. In cold regions, like Leh, hydroelectric power plants can operate only in summer months and during the rest of the year the population has to depend on diesel or other traditional sources to supplement their energy needs (Chandrasekharam *et al.* 2004). Oil and gas systems have similar problems as those of diesel systems.

The systems that appear to compete with geothermal cannot be sustained for long periods of time, and the unit cost of power from these systems cannot compete with electricity prices from geothermal power plants. It is then evident that small geothermal power plants have a large and sustainable market in developing countries. According to a survey reported by Lund and Boyd (1999), nearly 50 small geothermal power plants with a capacity of 5 MW$_e$ and below are successfully working in the world at present.

10.5   ADVANTAGES OF SMALL POWER PLANTS

There are several advantages for promoting small-scale geothermal power plants in rural areas (Entingh *et al.* 1999):

- Small power plants with generating capacity of 100–300 kW$_e$, along with its cooling system, can be transported easily on a container truck.
- Binary power plants can be operated with low-enthalpy fluids. Now that Kalina technology is in use, low-enthalpy geothermal resources can be utilized to the maximum extent to provide electricity to rural populations.
- The generating capacity of small power plants can be scaled up depending on the flow rate and fluid temperature. For example, as shown in Figure 10.3, wells with fluid temperature <150 °C und with fluid flow rate ⁓15 l/s can generate electric power anywhere between 50 kW$_e$ and

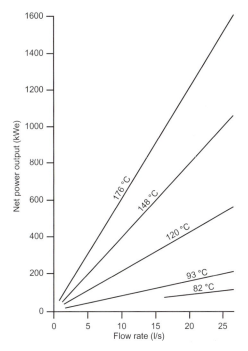

Figure 10.3.   Net power output for low-enthalpy geothermal fluids with different flow rates. Low-enthalpy systems with high flow rates can generate >100 kW$_e$. Clusters of several wells with this flow rate are best suited for rural areas (modified from Lund and Boyd 1999).

700 kW$_e$. Clusters of wells with different fluid temperatures and flow rates can also be used to plan to serve larger populations in rural areas.

- With current advancements in information technology, it is possible to operate and control small power plants remotely through satellites. This reduces human intervention and cost of man power to operate such systems. The Wineagle and Amedee geothermal power plants described above are operated remotely.
- Truck mounted drilling rigs can be used if the aquifers are located at shallow depths. Both the drilling and power plant equipment can be lifted by helicopters to remote areas where surface transportation is inadequate to move the equipment.
- If the composition of the fluids is clean, without causing any deteriorating effect, the expense of injection wells can be eliminated. The advantage here is that the flow rates are small and would not cause adverse effect on the soils or environment.
- Small power plants can have well heads near the plant itself thus reducing the cost of piping. In the case of clusters of wells, plastic, non-corrosive or carbon steel pipes can be used for connecting the pipes.
- Additional industrial applications can be planned near such plants. Greenhouse cultivation and/or dehydration units can be planned to use the available heat from the residual fluids from the heat exchangers of the power plant. This will create additional employment for the rural population and help them to lift their economic status.

10.6   COST OF SMALL POWER PLANTS

The reported costs of small geothermal power plants vary from 0.05 to 0.07 US$/kWh for units generating <5 MW$_e$ (Lund and Boyd 1999). A cost model developed by Entingh *et al.* (1994a, b) for systems with production capacity of <1 MW$_e$ is shown in Table 10.5. From the time this

Table 10.6.   Modeled cost of low-enthalpy small geothermal power plants.

Technical	
Resource temperature	120 °C
System net capacity	300 kWe
Number of wells	2
Capacity factor	0.8
Plant life time (years)	30
Rate of return on investment	12%/yr
kWh/yr produced	2.1 million
Capital costs (US$)	
Exploration (US$)	200000
Wells	325000
Field	94000
Power plant	659000
Total	1278000
Plant cost/installed kW	2200
Annual capital recovery cost	158650
O & M (US$/year)	
Field	32000
Plant	26000
Backup system	5000
Total/yr	63000

Source: Entingh *et al.* 1994b.

model was proposed until recently, several advancements in drilling technology, plant design, and working fluids have taken place. The present cost will be lower several fold compared to above mentioned 1999 estimates.

An analysis of the cost model shown in Table 10.5 indicates that, for current available technology, the plant cost can be reduced by adopting Kalina binary fluid technology, and exploration costs can be reduced, as described earlier, by utilizing available data from oil and groundwater wells. Costs can further be reduced because, geothermal power plants have the capacity to be on line >90% of the time by; therefore, backup system costs need not be included in the project cost. The unit cost of electricity can further be brought down if the power plants are located near the national grid network.

The modeled cost analysis given in Table 10.5 is applicable for geothermal fluids with inlet temperature ranges of 100 to 140 °C drawn from production well depths of about 200 to 1000 m and injection well depths of 200 to 500 m (Lund and Boyd 1999). Technical cost of such modeled systems at busbar is about 0.105 US$/kWh.

10.7   EXAMPLES OF SMALL POWER PLANTS

Lund and Boyd (1999) listed about 50 small geothermal power plants. In addition to the two plants described earlier, examples of a few important plants are given below.

10.7.1   *Chena low-enthalpy power plant, Alaska*

The recently installed geothermal power project at Chena geothermal springs site in Alaska is another excellent example that demonstrates the ability to utilize low-enthalpy geothermal sources for rural development.

Chena village in Alaska is remotely located and is connected by a paved road from Fairbanks. It is 53 km away from the nearest power grid. Electric power is supplied to the community through diesel at a unit cost of 0.30 US$ per kWh. The cost of electric power in rural Alaska is among the highest in the USA and is about 0.30 US$ per kWh sometimes touching as high as 1 US$. In 2005 US$ 365,000 was spent in diesel at a cost of US$ 2.5 per liter of diesel delivered in Chena.

However, this remote village (annual mean air temperature varying from 1 to $-6\,°C$) is gifted with a geothermal field. The discharge temperature of geothermal water in a bore well drilled to a depth of 300 m is $76\,°C$. This bore well, fitted with a down-hole pump, is able to supply fluids at the rate of 30 to 33 l/s. A production well with 25 cm ($10''$) diameter is located at a distance of about 4 km and supplies geothermal water at the above temperature.

This small-scale geothermal power project, Chena, Alaska, besides generating electric power at a cost of 0.05 US$ (and is expected to decrease further down to 0.01 US$/kWh), hosts several direct application projects reducing the unit cost of power further.

This project again demonstrates that downhole pump and binary technology are no longer barriers for developing small power projects in remote locations in both, developing and developed countries.

10.7.2 *TAD's enterprises binary plants, Nevada*

Two plants, installed in 1984 and 1987, are located in Nevada with generation capacity of 750 and 800 kW$_e$. Two wells supply geothermal water at $104\,°C$ with a flow rate of 60 l/s to these plants. Initially this plant was using Freon-114 as the binary fluid and due to non-availability of Freon-114, the plant was converted to use iso-pentane as the working fluid in 1998.

10.7.3 *Empire geothermal project, Nevada*

Located in the San Emidio desert, near Empire, Nevada, this project with four 1 MW$_e$ installed capacity turbines started operation in 1987. Two production wells with fluid temperature of $137\,°C$ were used. The water temperature dropped to $123\,°C$ due to cooling of the aquifer by injection wells. Dehydration units for onion and garlic in the same locality drew water from the same aquifer ($130\,°C$) causing a drop in the inlet water temperature of the geothermal plant. To maintain the production capacity of the plant, a third well with fluid temperature of $152\,°C$ was drilled in 1998. Thus, the output has been maintained at 3.85 MW$_e$ since 1998. The dehydration unit is also working successfully.

10.7.4 *Fang binary power plant, Thailand*

This plant was commissioned in 1989, with an installed capacity of 300 kW$_e$. The actual production varies from 150 to 250 kW$_e$. The water temperature is $116\,°C$ with a flow rate of about 8 l/s. The plant is online 97% and the unit cost of electricity varies between 0.06 and 0.08 US$/kWh. Though this appears expensive, it is much cheaper compared to the diesel based power that costs about 0.23 US$/kWh. The outlet fluid is being used for cold storage, crop drying and a spa.

10.7.5 *Nagqu binary plant, Tibet*

This plant is located at an elevation of >4000 m a.s.l. in Tibet. With an installed capacity of 1.3 MW$_e$, this plant is producing 1 MW$_e$, with a fluid flow rate of 69 l/s and a temperature of about $100\,°C$. This is a typical, stand-alone, rural and remotely based plant which is supplying electricity to nearly 20,000 people. Before the installation of this plant, diesel generators were used, with 4 to 5 hours of electricity supply with high unit cost. This plant has transformed the socio-economic status of the Tibetan population.

10.7.6   *Tu Chang binary power plant, Taiwan*

This 300 kW$_e$ plant is located in Taiwan and was commissioned in 1987. The plant is operated using 130 °C geothermal fluids drawn from a 500 m deep well. This plant is owned and operated by the Industrial Technology Research Institute and the power is sold to the Taiwan Power Company at 0.04 US$/kWh.

10.7.7   *Binary power plant in Copahue, Argentina*

This plant was built in 1988 with an installed capacity of 670 kW$_e$. This was the first geothermal power plant in South America, and is located at an altitude of about 2000 m a.s.l. in the Andes, in western Argentina. The plant runs on low pressure (6 t/h), 170 °C, and saturated steam. The plant was decommissioned in 1996 as the unit cost of electricity is much higher compared to natural gas. Here is an example where a geothermal power plant failed to compete with conventional energy but the prospects remain interesting: modern technology might make this plant economical in the future.

10.7.8   *Husavik, Kalina cycle binary power plant, Iceland*

Husavik is one the largest towns in northern Iceland with a population of 2500. Fishing is the main source of income for the inhabitants. Geothermal fields are located at a distance of about 25 km from the town. Initially (1967 to 1969) the geothermal waters were used for district heating, and subsequently during late 1990s, due to advancements made in binary plant technology, geothermal water from five wells drilled to a depth of about 450 m was utilized to generate electric power. The five wells are producing water with flow rates varying from 26 to 95 l/s, with temperatures varying from 115 to 128 °C. Electric power is generated, using a Kalina cycle-based binary power plant. The plant is generating 2 MW$_e$. Besides power, the geothermal energy from the field is being utilized for district heating, aquaculture, and recreation (warm bathing pools). The unit cost of electricity is about 0.13 US$/kWh for residential use and varies from 0.07 to 0.11 US$/kWh for industrial use (Hjartarson *et al.* 2002).

References

Afgan, N.H. and Carvalho, M.G.: Sustainability assessment of energy systems: An overview of current status. In: D. Chandrasekharam and J. Bundschuh (eds): *Geothermal energy for developing countries*. Balkema, Leiden, The Netherlands, 2002, pp.1–35.

Alam, M.A., Chandrasekharam, D., Vaselli, O., Capaccioni, B., Manetti, P. and Santo, A.B.: Petrology of the prehistoric lavas and dyke of the Barren Island, Andaman Sea, Indian Ocean. Proceedings Indian Acadamy of Science published in *Earth and Planetary Science* 113 (2004), pp.715–721.

Aldrich, M.J., Laughlin, A.W. and Gambil, D.T.: A review and augmentation of the Electric Power Research Institute report: Geothermal energy prospects for the next 50 Years. EPRI ER-611-SR; with comments on section 3, electrical energy conversion, by J.W. Tester, 1978.

Applegate, J.K. and Moens, T.A.: Geophysical logging case history of the Raft River geothermal system, Idaho. Los Alamos Scientific Laboratory informal report LA, 8252-MS, Los Alamos, CA, 1980.

Arnorsson, S.: Chemical equilibria in Iceland geothermal systems—Implications for chemical geothermometery investigations. *Geothermics* 12 (1983), pp.119–128.

Aspden, J.A., Kartawa, W., Aldiss, D.T., Djunuddin, A., Whandoyo, R., Diatma, D., Clarke, M.C.G. and Harahap, H.: The geology of the Padangsidempuan and Sibolga quadrangle. Geological Research and Development Centre, Department of Mines Indonesia, report, 1982.

Aunzo, Z., Laky, C., Steingrimsson, B., Bodvarsson, G.S., Lippmann, M.J., Truesdell, A., Escobar, C., Quintanilla, A. and Cuellar, G.: Pre-exploitation state of the Ahuachapán geothermal field, El Salvador. *Geothermics* 20:1/2 (1991), pp.1–22.

AWEA: Global Wind Energy Council press release: Global wind energy markets continue to boom—2006 another record year (Brussels, Belgium, February 2, 2007), http://www.awea.org/newsroom/index.html (accessed July 2007).

Barnes, D.F., Van der Plas, R. and Floor, W.: Tackling rural energy problem in developing countries. World Bank, Finance and Development, 34, Washington, DC, 1997, pp.11–15.

Battistellia, A., Yiheyisb, A., Calorec, C., Ferraginaa, C. and Abatnehb, W.: Reservoir engineering assessment of Dubti geothermal field, Northern Tendaho Rift, Ethiopia. *Geothermics* 31 (2002), pp.381–406.

Benito, F.A., Ogena, M.S. and Stimac, J.A.: Geothermal energy development in Philippines: Country update. *Proceedings World Geothermal Congress*, 24–29 April 2005, Antalya, Turkey, 2005.

Bertani, R.: World geothermal power generation in the period 2001–2005. *Geothermics* 34 (2005), pp.651–690.

Bhattacharji, S., Chatterjee, N., Wampler, J.M. and Gazi, M.: Mafic dykes in Deccan volcanics, indicator of Indian intraplate rifting, crustal extension and Deccan flood basalt volcanism. In: K.V. Subbarao (ed): *Volcanism*. John Wiley, New York, NY, 1994, pp.253–276.

Bibby, H.M., Caldwell, T.G., Davey, F.J. and Webb, T.H.: Geophysical evidence on the structure of the Taupo volcanic zone and its hydrothermal circulation. *J. Volcanology and Geothermal Research* 68 (1995), pp.29–58.

Bignall, G., Dorj, P., Batkhishing, B. and Tsuchiya, N.: Geothermal resources and development in Mongolia: Country update. *Proceedings World Geothermal Congress*, 24–29 April 2005, Antalya, Turkey, 2005.

Birkle, P. and Bundschuh, J.: High and low enthalpy geothermal resources and potentials. In: J. Bundschuh and G.E. Alvarado (eds): *Central America: Geology, resources and hazards, Volume 2*. Taylor and Francis/Balkema, Leiden, The Netherlands, 2007a, pp.705–776.

Birkle, P. and Bundschuh, J.: Hydrogeochemical and isotopic composition of geothermal fluids. In: J. Bundschuh and G.E. Alvarado (eds): *Central America: Geology, resources and hazards, Volume 2*. Taylor and Francis/Balkema, Leiden, The Netherlands, 2007b, pp.777–838.

Birkle, P. and Merkel, B.: Environmental impact by spill of geothermal fluids at the geothermal field of Los Azufres, Michoacán, Mexico. *Water, Air and Soil Pollution* 124:3/4 (2000), pp.371–410.

Birkle, P. and Merkel, B.: Mineralogical-chemical composition and environmental risk potential of the evaporation pond sediments at the geothermal field of Los Azufres, Michoacán, México. *Environmental Geology* 41 (2002), pp.583–592.

119

Biswas, S.K.: Regional tectonic framework, structure and evolution of western marginal basin of India. *Tectonophysics* 135 (1987), pp.305–327.

Björnsson, J., Helgason, T., Palmason, G., Stefansson, V., Jonatansson, H., Mariusson, J.M., Fridleifsson, I.B. and Thorsteinsson, L.: The potential role of geothermal energy and hydropower in the world energy scenario in year 2020. Paper 3.1.07, *Proceedings of the 17k WEC Congress*, Houston (Texas), Volume 5, 1988, pp.69–87.

Bloomster, C.H. and Maeder, P.F.: Economic considerations. In: R. DiPippo, H.E. Khalifa and D.J. Ryley (eds): *Source book on the production of electricity from geothermal energy*. US Department of Energy, Washington, DC, 1980, pp.682–713.

Brown, K.: Environmental aspects of geothermal development. World Geothermal Congress, Pre-Congress Courses, International Geothermal Association (ed), Pisa, Italy, 1995.

Bruno, P.P.G., Paoletti, V., Grimaldi, M. and Rapolla, A.: Geophysical exploration for geothermal low enthalpy resources in Lipari Island, Italy. *J. Volcanology Geothermal Research* 98 (2000), pp.173–188.

Bundschuh, J., Alvarado, G.E., Rodriguez, J.A., Roldán, A.R., Palma, J.C., Zuñiga, A., Reyes, E. and Castillo, G.: Resources and policy of geothermal energy in the Central America. In: D. Chandrasekharam and J. Bundschuh (eds): Geothermal energy for developing countries. A.A. Balkema, Leiden, The Netherlands, 2002, pp.313–364.

Bundschuh, J., Birkle, P., Aaheim, A. and Alvarado, G.E.: Geothermal resources for development—valuation, present use and future opportunities. In: J. Bundschuh and G.E. Alvarado (eds): *Central America: Geology, resources and hazards, Volume 2*. Taylor and Francis/Balkema, Leiden, The Netherlands, 2007a, pp.869–894.

Bundschuh, J., Winograd, M., Day, M. and Alvarado, G.E.: Geographical, social, economic, and environmental framework and developments. In: J. Bundschuh and G.E. Alvarado (eds): *Central America: Geology, resources and hazards, Volume 1*. Taylor and Francis/Balkema, Leiden, The Netherlands, 2007b, pp.1–52.

Caicedo, A.A.: Currrent status of geothermal activities in Guatemala. In: R.L. Miller, G. Escalante, J.A. Reinemund and M.J. Bergin (eds): *Energy and mineral potential of the Central American-Caribbean regions*. Circum-Pacific Council for Energy and Mineral Resources, Earth Science Series, Vol. 16, Springer-Verlag, Berlin, Germany, 1995, pp.247–255.

California Energy Company, Inc.: Geology and temperature distribution of the Momotombo geothermal field Nicaragua, unpublished, 1979.

Cataldi, R., Stefani, G. and Tongiorgi, M.: Geology of Larderello region (Tuscany): contribution to the study of geothermal basins. *Proceedings Spoleto Meeting on Nuclear Geology in Geothermal Areas*, Laboratorio Geologia Nuclearc, Pisa, 1963, pp.235–265.

Cataldi, R., Stefani, G.C. and Tongiorgi, E.: Geology of Larderello region (Tuscany); contribution to the study of the geothermal basin. In: E. Tongiorgi (ed): *Nuclear geology on geothermal areas*. Proceedings International Symposium, Spoleto, Italy, 1963, pp.235–265.

Central Intelligence Agency (CIA): *World Fact Book 2007*. Central Intelligence Agency, Washington, DC, https://www.cia.gov/library/publications/the-world-factbook (accessed July 2007).

Chadha, R.K.: Geological contacts, thermal srings and earthquakes in peninsular India. *Tectonophysics* 213 (1992), pp.367–374.

Chandrasekharam, D.: Geothermal energy resources of India—Country update. *Proceedings World Geothermal Congress 2000*, Kyushu-Tohoku, Japan, May 28–June 10, 2000, pp.133–145.

Chandrasekharam, D.: Geothermal energy resources of India: Past and the present. *Proceedings World Geothermal Congress*, 24–29 April 2005, Antalya, Turkey, 2005.

Chandrasekharam, D.: Structure and evolution of the western continental margin of India deduced from gravity, seismic, geomagnetic and geochronological studies. *Physics of the Earth and Planetary Interiors* 41 (1985), pp.186–198.

Chandrasekharam, D.: HDR prospects of Himalaya geothermal province. In: K. Popovski and B. Sanner (eds): *Proceedings International Geothermal Days*, 17–22 September 2001, Bad Urach, Germany, 2001a, pp.315–320.

Chandrasekharam, D.: Opportunities for small scale geothermal power projects in India. *Industrial Products Finder* 29 (2001b), pp.203–205.

Chandrasekharam, D.: Use of geothermal energy for food processing—Indian status. *Geo-Heat Centre Quarterly Bulletin* 22 (2001c), pp.8–11.

Chandrasekharam, D.: Industrial applications of geothermal energy. *Industrial Products Finder* 23 (1995), pp.223–225.

Chandrasekharam, D. and Antu, M.C.: Geochemistry of Tattapani thermal springs, Madhya Pradesh, India: Field and experimental investigations. *Geothermics* 24 (1995), pp.553–559.

Chandrasekharam, D. and Ayaz, M.A.: Direct utilization of geothermal energy resources—NW Himalayas, India. *Geothermal Resources Council Transactions* 27 (2003), 81–83.

Chandrasekharam, D. and Bundschuh, J.: Geothermal energy resources for developing countries. A.A. Balkema, Leiden, The Netherlands, 2002.

Chandrasekharam, D. and Chandrasekhar, V.: Enhanced geothermal resources: Indian scenario. *Geothermal Resources Council Transactions* 31 (2007), pp.271–273.

Chandrasekharam, D. and Chandrasekhar, V.: Geothermal energy resources of India: Ongoing and future developments. *American Association of Petroleum Geologists Conference Proceedings*, 5–8 November 2006, Perth, Australia, 2006.

Chandrasekharam, D. and Jayaprakash, S.J.: Geothermal energy assessment: Bugga and Manuguru thermal springs, Godavari valley, Andhra Pradesh. *Geothermal Resources Bulletin* 25 (1996), pp.19–21.

Chandrasekharam, D. and Prasad, S.R.: Geothermal system in Tapi rift basin, northern Deccan province, India. In: G.B. Arehart and J.R. Hulston (eds): *Proceedings 9ᵗʰ Water-Rock Interaction*. A.A. Balkema, Leiden, The Netherlands, 1998, pp.667–670.

Chandrasekharam, D., Alam, M.A. and Minissale, A.: Geothermal resources potential of Himachal Pradesh. In: I.B. Fridleifsson and E.T. Eliasson (eds): *Multiple integrated use of geothermal resources*. International Geothermal Conference, Reykjavik, Iceland, 2003, pp.15–19.

Chandrasekharam, D., Alam, M.A. and Minissale, A.: Geothermal potential of the Ladakh region, NW Himalayas, India. *Proceedings International Conference on Sustainable Habitat for cold climates*, 17–18 September 2005, Leh, Ladakh, India, 2005, pp.1–5.

Chandrasekharam, D., Alam, M.A. and Minissale, A.: Thermal discharges at Manikaran, Himachal Pradesh, India. *Proceedings World Geothermal Congress*, 24–29 April 2005, Antalya, Turkey, 2005.

Chandrasekharam, D., Ramanathan, A. and Selvakumar, R.L.: Thermal springs in the Precambrian crystallines of western continental margin of India—field and experimental results. In: Y. Kharaka and A. Maest (eds): *Proceedings 7ᵗʰ Water-Rock Interaction*. A.A. Balkema, Leiden, The Netherlands, 1992, pp.1271–1274.

Chandrasekharam, D., Ramesh, R. and Balasubramanian, J.: Geochemistry, oxygen and hydrogen isotope ratios of thermal springs of western continental margin of India—field and experimental results. In: D.L. Miles (ed): *Proceedings 6ᵗʰ Water-Rock Interaction*. A.A. Balkema, Leiden, The Netherlands, 1989, pp.149–154.

Chen, C.T.A. and Marshall, W.L.: Amorphous silica solibilities—IV. Behavior in pure water and aqueous sodium chloride, sodium sulfate, magnesium chloride and magnesium sulfate up to 350 °C. *Geochimica et Cosmochimica Acta* 46 (1982), pp.279–287.

Clarke, M.C.G., Woodhall, D.G., Allen, D. and Darling, G.: Geological, volcanological and hydrogeological controls on the occurrence of geothermal activity in the area surrounding Lake Naivasha Keuya. British Geological Survey report to the Ministry of Energy, Nairobi, Kenya, 1990.

Clauser, C. and Huenges, E.: Thermal conductivity of rocks and minerals. In: *Rock physics and phase relations—A handbook of physical constants*. Amer. Geophy. Union Reference Shelf, 3, 1995, pp.105–127.

Clemente, W. and Abrigo, F.L.V.: The Bulalo geothermal field, Philipphines: Reservoir characteristics and response to production. *Geothermics* 22 (1993), pp.381–394.

Combs, J. and Muffler, L.J.P.: Exploration for geothermal resources. In: P. Kruger and C. Otte (eds): *Geothermal Energy*. Stanford, CA, 1973, pp.95–128.

Cooper, M.A., Herbert, R. and Hill, G.S.: The structural evolution of Triassic intermontane basins in north-eastern Thailand. *Proceedings International Symposium on Intermonte Basins: Geology, and Resources*, Chiang Mai, Thailand, 1989, pp.231–242.

Cordon, U.J. Momotombo field models at six stages in time. *Geothermal Resources Council Transactions* 4 (1980), pp.443–446.

Craig, H.: Standard for reporting concentrations of deuterium and oxygen-18 in natural waters. *Science* 133 (1961), pp.1833–1934.

Curray, J.R., Moore, D.G., Lawver, L.A., Emmel, F.J., Raitt, R.W., Henry, M. and Kieckhefer, R.M.: Tectonics of the Andaman Sea and Burma. In: J. Watkins, L. Montadert and P.W. Dickenson (eds): *Geological and geophysical investigations of continental margins*. CCOP–SEATAR Meeting, Bandung, American Association of Petroleum Geology Memoir 29, 1979, pp.189–198.

Curray, J.R., Shot, Jr., G.G., Raitt, R.W. and Henry, M.: Seismic refraction and reflection studies of crustal structure of the eastern Sunda and western Banda arcs. *J. Geophysical Research* 82 (1977), pp.2479–2489.

Dart, R.H. and Whitbeck, J.F.: Binary heat exchangers. In: R. DiPippo, H.E. Khalifa and D.J. Ryley (eds): *Source book on the production of electricity from geothermal energy*. US Department of Energy, Washington, DC, 1980, pp.379–412.

DeMets, C., Gordon, R.G., Argus, D.F. and Stein, D.F.: Current plate motion. *Geophysical Journal International* 101 (1990), pp.425–478.

de Luga, P.P., LaTerra, E.F., Kriegshauser, B. and Fontes, S.L.: Magnetotelluric studies of the Caldas Novas geothermal reservoir, Brazil. *J. Applied Geophysics* 49 (2002), pp.33–46.

DiPippo, R.: *Geothermal power plants: Principles, applications and case studies.* Elsevier, New York, NY, 2005.

Dor Ji and Ping, Z.: Characteristics and genesis of the Yangbajing geothermal field, Tibet. *Proceedings World Geothermal Congress 2000*, May 28–June 10, Kyushu-Tohoku, Japan, pp.1083–1088.

Dorj, P.: Design of a small geothermal heating systems and power generators for rural consumers in Mongolia. Report 3, United Nations Training Programme, Iceland, 2001, pp.27–57.

Downey, W.S., Kellett, R.J., Smith, I.E.M., Price, R.C. and Stewart, R.B.: New paleomagnetic evidence for the recent eruptive activity of Mt. Taranki, New Zealand. *J. Volcanology and Geothermal Research* 60 (1994), pp.15–27.

Einarsson, S.: Study of the temperature distribution in the geothermal reservoir at Momotombo and its implications. United Nations Development Program (UNDP), Geothermal Resources Development, Nic/74/003, unpublished, 1977.

Electroconsult: Momotombo geothermal field feasibility report, GNI-D-3820. Internal report, 1977.

Ellis, A.J. and Mahon, W.A.: *Chemistry and geothermal systems.* Academy Press, New York, NY, 1977.

Endeshaw, A.: Current status (1987) of geothermal exploration in Ethiopia. *Geothermics* 17 (1988), pp.477–488.

Energy Information Administration (EIA): International energy outlook 2007. Energy Information Administration, Office of Integrated Analysis and Forecasting, US Department of Energy, Washington, DC, DOE/EIA-0484(2007), 2007, http://www.eia.doe.gov/oiaf/ieo/index.html (accessed July 2007).

Entingh, D.J., Easwaran, E. and McLarty, L.: Small geothermal electrical systems for remote powering. *Geothermal Resources Council Transactions* 18 (1994a), pp.39–45.

Entingh, D.J., Easwaran, E. and McLarty, L.: Small geothermal electrical systems for remote power. *Geothermal Resources Council Bulletin* 23 (1994b), pp.331–338.

Faure, G.: *Principles of isotope geology.* John Wiely and Sons, New York, NY, 1986.

Fauzi, A., Suryadarma, Soemarinda, S. and Siahaan, E.E.: The role of Pertamina in geothermal development in Indonesia. Proceedings World Geothermal Congress, 24–29 April 2005, Antalya, Turkey, pp.1–6.

Fauzil, A., Bahri, S. and Akuanbatin, H.: Geothermal development in Indonesia: An overview of industry status and future growth. *Proceedings World Geothermal Congress 2000*, May 28–June 10, Kyushu-Tohoku, Japan, 2000, pp.1109–1114.

Fournier, R.O. and Rowe, J.J.: Estimation of underground temperatures from silica content of water from hot springs and wet-steam wells. *American Journal of Science* 264 (1966), pp.685–697.

Fournier, R.O.: Silica in thermal waters: Laboratory and field investigations. *Proceedings International Symposium on hydrogeochemistry and biogeochemistry*, 1, Tokyo, 1973, pp.122–139.

Fournier, R.O. and Potter, R.W.: An equation correlating the solubility of quartz in water from 25 °C to 900 °C at pressure up to 10,000 bars. *Geochimica et Cosmochimca Acta* 46 (1982), pp.1969–1974.

Fournier, R.O.: A method for calculating quartz solubilities in aqueous sodium chloride solutions. *Geochimica et Cosmochimica Acta* 47 (1983), pp.579–586.

Fournier, R.O. and Marshall, W.L.: Calculations of amorphous silica solubilities at 25 to 300 °C and apparent cation hydration numbers in aqueous salt solutions using the concept of effective density of water. *Geochimica et Cosmochimica Acta* 47 (1983), pp.587–596.

Fournier, R.O.: The behavior of silica in hydrothermal solutions. In: B.R. Berger and P.M. Bethke (eds): *Reviews in economic geology, volume 2.* The Economic Geology Publishing Company, Littleton, CO, 1985, pp.45–61.

Gamble, J.A., Smith, I.E.M., McCulloch, M.T., Graham, I.J. and Kokelaar, B.P.: The geochemistry and petrogenesis of basalts from the Taupo volcanic zone and Kermadec island arc, SW Pacific. *J. Volcanology and Geothermal Research* 54 (1993), pp.265–290.

Gawell, K., Reed, M. and Wright, P.M.: Preliminary report: Geothermal energy, the potential for clean power from the earth. Issued April 7, 1999 by the Geothermal Energy Association, Washington, DC, 1999.

Gendenjamts, O.E.: Geochemical study of Mongolia hot springs. *Proceedings World Geothermal Congress*, 24–29 April 2005, Antalya, Turkey, 2005.

GeothermEx: New geothermal site identification and qualification. Consultant report for California Energy Commission, Richmond, CA, 2004.

Giggenbach, W.F.: Isotopic composition of waters from Broadlands geothermal field (New Zealand). *New Zeland Journal of Science* 14 (1971), pp.959–970.

Giggenbach, W.F.: The isotopic composition of waters from El Tatio geothermal field, northern Chile. *Geochimica et Cosmochimca Acta* 42 (1978), pp.143–161.

Giggenbach, W.F. and Lyon, G.L.: The chemical and isotopic compositions of water and gas discharges from the Ngawha geothermal field. New Zeland DSIR geothermal report, 1977, pp.30–37.

Giggenbach, W.F. and Stewart, M.K.: Processes controlling the isotopic composition of steam and water discharges from steam vents and steam-heated pools in geothermal areas. *Geothermics* 11 (1982), pp.71–80.

Giggenbach, W.F., Gonfiantini, R. and Panichi, C.: Geothermal systems. In: *Guide book on nuclear techniques in hydrology.* IAEA, Vienna, Tech. Rep. 91, 1983, pp.359–379.

Giggenbach, W.F.: Geothermal solute equilibria: Derivation of Na-K-Mg-Ca geoindicators. *Geochimica et Cosmochimica Acta* 52 (1988), pp.2749–2765.

Goff, F., Goff, S.J., Shevenell, L., Truesdell, A.H., Musgrave, J., Rüfenacht, H. and Flores, W.: Exploration drilling and reservoir model of the Platanares geothermal system, Honduras, Central America. *J. Volcanology and Geothermal Research* 45 (1991), pp.101–123.

Goldsmith, L.H.: Regional and local geologic structure of the Momotombo field, Nicaragua. *Geothermal Resources Council Transactions* 4 (1980), pp.125–128.

González, E.P., Torres, V.R. and Birkle, P.: Plio-Pleistocene volcanic history of the Ahuachapán geothermal system, El Salvador: The Concepción de Ataco caldera. *Geothermics* 26:5/6 (1997), pp.555–575.

Gopalan, K., Trivedi, J.R., Mayor, S.S., Patel, P.P. and Patel, S.G.: Rb-Sr ages of Godhra and related granites, Gujarat, India. Proceeding of the Indian Academy of Science, published in *Earth and Planetary Science* 88A (1979), pp.7–17.

Grimaud, D., Huang, S., Michard, G. and Zheng, K.: Chemical study of geothermal waters of central Tibet (China). *Geothermics* 14 (1995), pp.35–48.

Gunderson, R., Ganefianto, N., Riedel, K., Sirad-Azwar, L. and Suleiman, S.: Exploration results in the Sarulla block, north Sumatra, Indonesia. *Proceedings World Geothermal Congress 2000*, May 28–June 10, Kyushu-Tohoku, Japan, 2000, pp.1183–1188.

Hance, C.N.: Factors affecting costs of geothermal power development. Geothermal Energy Association, US Department Energy, Washington, DC, 2005.

Harinarayana, T., Abdul Azeez, K.K., Murthy, D.N., Veeraswamy, Eknath Rao, S.P., Manoj, C. and Naganjaneyulu, K.: Exploration of geothermal structure in Puga geothermal field, Ladakh Himalayas, India by magnetotelluric studies. *J. Applied Geophysics* 58 (2006), pp.280–295.

Haris, N., Vance, D. and Ayres, M.: From sediment to granite: timescales of anatexis in the upper crust. *Chemical Geology* 162 (2000), pp.155–167.

Harrison, T.M., Groove, M., Lovera, O.M. and Catlos, E.J.: A model for the origin of Himalayan antexis and inverted metamorphism. *J. Geophysical Research* 103 (1998), pp.27,017–27,032.

Harrison, T.M., Grove, M., McKeegan, K.D., Coath, C.D., Lovera, O.M. and Le Foort, P.: Origin and episodic emplacement of the Manaslu intrusive complex, Central Himalayas. *J. Petrology* 40 (1999), pp.3–19.

Henley, R.W., Truesdell, A.H. and Barton Jr., P.B.: Fluid-mineral equilibria in hydrothermal systems. *Reviews in Economic Geology, volume 1*. The Economic Geology Publishing Company, Littleton, CO, 1985.

Hjartarson, H., Maack, R. and Johannesson, S.: Husavk Energy: Multiple use of geothermal energy. Thermal project no: GE 321/98/IS/DK, National Energy Authority, Reykjavík, Iceland, 2002.

Hochstein, M.P. and Regenauer-Lieb, K.: Heat generation associated with collision of two plates: the Himalayan geothermal belt. *J. Volcanology and Geothermal Research* 83 (1998), pp.75–92.

Hoke, L., Lamb, S., Hilton, D.R. and Poreda, R.J.: Southern limit of mantle-derived geothermal helium emissions in Tibet: implications for lithospheric structure. *Earth and Planetary Science Letters* 180 (2000), pp.297–308.

Hooper, P.R.: The timing of crustal extension and the eruption of continental flood basalts. *Nature* 345 (1990), pp.246–249.

Ibrahim, R.F., Fauzi, A. and Suryadarma: The progress of geothermal energy resources activities in Indonesia. *Proceedings World Geothermal Congress*, 24–29 April 2005, Antalya, Turkey, 2005.

International Heat Flow Commission: The global heat flow database. International Heat Flow Commission; http://www.heatflow.und.edu/ (accessed July 2007).

Jacobs, H.R. and Boehm, R.F.: Direct contact binary cycles. In: R. DiPippo, H.E. Khalifa and D.J. Ryley (eds): *Source book on the production of electricity from geothermal energy.* US Department of Energy, Washington, DC, 1980, pp.413–471.

Jain, S.C., Nair, K.K.K. and Yedekar, D.B.: Geology of the Son-Narmada-Tapti lineament zone in central India. *Geological Survey of India, Special Publication* 10, 1995, pp.1–154.

Joga Rao, M.V., Rao, A.P., Midha, R.K., Padmanabhan, K. and Kesavamani, K.: Results of geophysical survey in the Tattapani hot springs area, Sarguja district, Madhya Pradesh, *Geological Survey of India Records* 115 (1986), pp.66–83.

Jonsson, M.: *Advanced power cycles with mixtures as the working fluid.* PhD Thesis, Royal Institute of Technology, Stockholm, Sweden, 2003.

Kalina, A.I.: Combined cycle and waste heat recovery power systems based on a novel thermodynamic energy cycle utilizing low-temperature heat for power generation. *Proceedings of the 1983 Joint Power Generation Conference*, Indianapolis, IN, 1983. ASME Paper No. 83-JPGC-GT-3, American Society of Mechanical Engineers, New York, NY, 1983, pp.1–5.

Kalina, A.I.: Combined cycle system with novel bottoming cycle. *J. Engineering for Gas Turbines and Power* 106 (1984), pp.737–742.

Kalina, A.I. and Leibowitz, H.M.: Applying Kalina technology to a bottoming cycle for utility combined cycles. *Proceedings of the Gas Turbine Conference and Exhibition*, 31 May–June 4, 1987, Anaheim, CA, ASME Paper No. 87-GT-35, 1987.

Kalina, A.I. and Leibowitz, H.M.: Application of the Kalina cycle technology to geothermal power generation. *Geothermal Resources Council Transactions* 13 (1989), pp.605–611.

Kaila, K.L., Rao, I.B.P., Koteswar Rao, P., Madhava Rao, Krishna, V.G. and Sridhar, A.R.: DSS studies over Deccan trap along the Thuadara-Sendhwa-Sindad profile across Narmada-Son lineament, India. In: R.F. Mereu, S. Muller and D.M. Fauntain (eds): *Properties of earth's lower crust.* American Geophysical Union Monograph 51, 1981, pp.127–141.

Kasameyer, R.: Brief guidelines for the development of inputs to CCTS from the technology working group. Working draft, Lawrence Livermore Laboratory, Livermore, CA, 1997.

Kearey, P. and HongBing, W: Geothermal fields of China. *J. Volcanology and Geothermal Research* 56 (1993), pp.415–428.

Laky, C., Lippmann, M.J., Bodvarsson, G.S., Retana, M. and Cuellar, G.: Hydrogeological model of the Ahuachapán geothermal field, El Salvador. *Proc. 14th Workshop on Geothermal Reservoir Engineering*, Stanford University, CA, 1989, pp.265–272.

Lazzeri, L.: Application of Kalina cycle as bottoming cycle for existing geothermal plants. *Proceedings of the Florence World Energy Research Symposium*, July 30–August 1, 1997, Florence, Italy, 1997, pp.389–396.

Le Fort, P. and Rai, S.M.: Pre-Tertiary felsic magmatism of the Nepal Himalaya: recycling of continental crust. *J. Asian Earth Sciences* 17 (1999), pp.607–628.

Lima, E.M.L., Palma, J. and Roldán Manzo, A.R.: Geothermal Guatemala—Past, present and future development of geothermal energy in Guatemala, Guatemala. *Geothermal Resources Council Bulletin* (May/June 2003), 2003, pp.117–121.

Lkhagvadorj, I. and Tseesuren, B.: Geothermal energy resources, present utilization and future developments in Mongolia. *Proceedings World Geothermal Congress*, 24–29 April 2005, Antalya, Turkey, 2005.

Lund, J.: 100 years of geothermal power production. *Geo Heat Centre Bulletin* 25 (2004), pp.11–19.

Lund, J. and Boyd, T.: Small geothermal power projects examples. *Geo Heat Centre Bulletin* 20 (1999), pp.9–26.

Lund, J.W., Freeston, D.H. and Boyd, T.L.: Direct application of geothermal energy: 2005 Worldwide review. *Geothermics* 34 (2005), pp.691–727.

Makovsky, Y., Klemperer, S.L., Ratschbacher, L. and Alsdorf, D.: Midcrustal reflector on INDEPTH wide-angle profiles: an ophiolitic slab beneath the India—Asia suture in southern Tibet? *Tectonics* 18 (1999), pp.793–808.

Makundi, J.S. and Kifua, G.M.: Geothermal features of the Mbeya prospects in Tanzania. *Geothermal Resources Council Transaction* 9 (1985), pp.451–454.

Martínez Tiffer, E., Arcia Lacayo, R. and Sabatino, G.: Geothermal development in Nicaragua. *Geothermics* 17:2/3 (1988), pp.333–354.

McCaffrey, R.: Oblique plate convergence, slip vectors, and fore-arc deformation. *J. Geophysical Research* 97 (1992), pp.8905–8915.

McCarthy, A.J. and Elders, C.F.: Caenozoic deformation in Sumatra oblique subduction and the development of the Sumatran fault system. In: A.J. Fraser and S.J. Matthews (eds): *Petroleum geology of SE Asia.* Geological Society of London Special Publication 126, 1997, pp.355–363.

McKenzie, W.F. and Truesdell, A.H.: Geothermal reservoir temperatures estimated from the oxygen isotope compositions of dissolved sulfate and water from hot springs and shallow drillholes. *Geothermics* 5 (1977), pp.51–61.

Minissale, A.: The Larderello geothermal field: A review. *Earth Science Reviews* 31 (1991), pp.133–151.

Minissale, A., Chandrasekharam, D., Vaselli, O., Magro, G., Tassi, F., Pansini, G.L. and Bhramhabut, A.: Geochemistry, geothermics and relationship to active tectonics of Gujarat and Rajasthan thermal discharge, India. *J. Volcanology and Geothermal Research* 127 (2003), pp.19–32.

Minissale, A., Vaselli, O., Chandrasekharam, D., Magro, G., Tassi, F. and Casiglia, A.: Origin and evolution of "intracratonic" thermal fluids from central western peninsular India. *Earth and Planetary Science Letters* 181 (2000), pp.377–394.

Minster, J.B. and Jordan, T.H.: Present-day plate motions. *J. Geophysical Research* 83 (1978), pp.5331–5334.

MIT: The future of geothermal energy—Impact of enhanced geothermal systems (EGS) on the United States in the 21st century. An assessment by an MIT-led interdisciplinary panel, Massachusetts Institute of Technology, 2006, http://geothermal.inel.gov and http://www1.eere.energy.gov/geothermal/egs_technology.html (accessed July 2007).

Mitchell, A.H.G.: Cretaceous-Cenozoic tectonic events in the western Myanmar (Burma)-Assam region. *J. Geological Society, London* 150 (1993), pp.11,089–11,102.

Montalvo, F., D'Amore, F., Tenorio, J. and Marínez, M.: Twenty years of exploitation at Ahuachapán geothermal field: An assessment of the chemical and physical reservoir parameters. *Proc. 22nd Workshop on Geothermal Reservoir Engineering*, Stanford University, CA, 1997, pp.45–53.

Moya, P., Manieri, A. and Yock, A.: Development of geothermal energy in Costa Rica. In: D. Chandrasekharam and J. Bundschuh (eds): *Geothermal energy for developing countries*. Taylor and Francis/Balkema, Leiden, The Netherlands, 2002, pp.365–384.

Moya, P.: Overview of the Miravalles, Las Pailas and Boriquen geothermal zones. 20th meeting of the Panel de Consultores de Miravalles, Las Pailas y Borinquen, March 2006, Instituto Costarricense de Electricidad, San José, Costa Rica, 2006.

Naqvi, S.M., Divakar, Rao, V. and Han Narain: The primitive crust: Evidence from the Indian shield. *Precambrian Research* 6 (1978): pp.323–345.

Nathenson, M., Nehring, N.L., Crosthwaite, E.G., Harmon, R.S., Janik, C. and Borthwick, J.: Chemical and light stable isotope characteristics of waters from the Raft river geothermal area and environs, Cassia county, Idaho; Box Elder county, Utah. *Geothermics* 11 (1982), pp.215–237.

Negi, J.G., Agrawal, P.K., Singh, A.P. and Pandey, O.P.: Bombay gravity high and eruption of Deccan flood basalts (India) from a shallow secondary plume. *Tectonophysics* 206 (1992), pp.341–350.

Nesbit, B.E.: Electrical resistivities of crustal fluids. *J. Geophysical Research* 98 (1993), pp.4301–4310.

Noble, S.R., Searle, M.P. and Walker, C.B.: Age and tectonic significance of Permian granites in western Zanskar, High Himalaya. *J. Geology* 109 (2001), pp.127–135.

Omenda, P.A.: The geology and structural controls of the Olkaria geothermal systems, Kenya. *Geothermics* 27 (1998), pp.55–74.

Oskooi, B.: 1-D interpretation of the magnetotelluric data from Travale geothermal field in Italy. *J. Earth and Space Physics* 32 (2006), pp.1–16.

Page, B.G.N., Bennet, J.D., Cameron, N.R., Bridge, D., McC Jeffery, D.H., Keats, W. and Thaib, J.: A review of the main structural and magmatic features of northern Sumatra. *J. Geological Society London* 136 (1979), pp.569–579.

Palma, J. and García, O.: Updated status of the geothermal development in Guatemala. *Proceedings World Geothermal Congress*, 24–29 April 2005, Antalya, Turkey, 2005.

Pornuevo, J.B.: Lithologic map space of Southern Negros geothermal project. Unpublished internal report, Indonesia, 1984.

Porras, E.A., Tanaka T., Fujii, H. and Itoi, R.: Numerical modeling of the Momotombo geothermal system, Nicaragua. *Proceedings World Geothermal Congress*, 24–29 April 2005, Antalya, Turkey, 2005.

Rae, A.J., Cooke, D.R., Phillips, D. and Zaide-Delfin, M.: The nature of magmatism at Palinpinon geothermal feld, Negros Island, Philippines: implications for geothermal activity and regional tectonics. *J. Volcanology and Geothermal Research* 129 (2004), pp.321–342.

Raju, A.T.R. and Srinivasan, S.: Cambay basin petroleum habitat. In: S.K. Biswas (ed): *Proceedings Petroleferous Basins of India*, vol. 2, 1993, pp.33–78.

Ramanathan, A. and Chandrasekharam, D.: Geochemistry of Rajapur and Puttur thermal springs, west coast of India. *J. Geological Society India* 49 (1997), pp.559–565.

Ramingwong, T. and Lertsrimongkol, S.: Update on geothermal development in Thailand. *Proceedings World Geothermal Congress*, Florence, 1995, pp.337–340.

Ravi Shanker: Thermal and crustal structure of "SONATA". A zone of mid continental rifting in Indian shield. *J. Geological Society of India* 37 (1991), pp.211–220.

Reyes, A.G and Jongens, R.: Tectonic settings of low enthalpy geothermal systems in New Zealand: An overview. *Proceedings World Geothermal Congress*, 24–29 April 2005, Antalya, Turkey, 2005.

Riaroh, D. and Okoth, W.: The geothermal fields of the Kenya rift. *Tectonophysics* 236 (1994), pp.17–130.

Ripperda, M., Bodvarsson, G.S., Cuellar, G., Escobar, C. and Lippmann, M.J.: An exploitation model and performance predictions for the Ahuachapán geothermal field, El Salvador. *Proceedings 15th Workshop on Geothermal Reservoir Engineering*, Stanford University, CA, 1990, pp.147–153.

Ripperda, M., Bodvarsson, G.S., Lippmann, M.J., Cuellar, G. and Escobar, C.: An exploitation model and performance predictions for the Ahuachapán geothermal field, El Salvador. *Geothermics* 20:4 (1991), pp.181–196.

Risk, G.F.: Detection of buried zones of fissured rock geothermal fields using resistivity anisotropy measurements. *Proceedings* 2nd *United Nations Symposium*, San Francisco, CA, 1976, pp.1191–1198.

Rodolfo, K.S.: Bathymetry and marine geology of the Andaman basin and tectonic implications for Southeast Asia. *Geological Society of America Bulletin* 80 (1969), pp.1203–1230.

Roldán, M.A.R.: Investigaciones geoquímicas realizadas en los campos geotérmicos de Zunil y Amatitlán, Guatemala. International Atomic Energy Agency (IAEA) TECDOC-641, Vienna, 1992, pp.279–305.

Roldán, M.A.R.: Geothermal power development in Guatemala 2000–2005. *Proceedings World Geothermal Congress*, 24–29 April 2005, Antalya, Turkey, 2005.

Rosell, J.B. and Zaide-Delfin, M.C.: Resource potential of the southern Leyte geothermal prospect, Philippines: A geologic evaluation. *Proceedings World Geothermal Congress*, 24–29 April 2005, Antalya, Turkey, 2005.

Rowley, J.C.: Worldwide geothermal resources. In: L.H. Edwards, G.V. Chilingar, H.H. Rieke and W.H. Fertl (eds): *Handbook of geothermal energy*. Gulf Publ. Co., Houston, TX, 1982, pp.44–176.

Rybach, L., Haenel, R. and Stegena, L.: *Handbook of terrestrial heat-flow density determination*. Kluwer Academic Publishers, London, 1988.

Sachan, H.K.: Cooling history of subduction related granite from the Indus suture zone, Ladakh, India: evidence from fluid inclusions. *Lithos* 38 (1996), pp.81–92.

Sakungo, F.K.: Geothermal resources of Zambia. *Geothermics* 17 (1988), pp.503–514.

Sanner, B. and Anderson, O.: Drilling methods for shallow geothermal installations. In: K. Popovski and B. Sanner (eds): *Proceedings International Summer School on direct application of geothermal energy*, Bad Urach, Germany, 2002, pp.25–43.

Sanyal, S., Kitz, K. and Glaspey, D.: Optimization of power generation from moderate temperature geothermal systems—A case history. *Geothermal Resources Council Transactions* 29 (2005), pp.627–633.

Schneider, D.A., Edwards, M.A., Zeitler, P.K. and Coath, C.D.: Mazeno Pass pluton and Jutial pluton, Pakistan Himalaya: age and implications for entrapment mechanism of two granites in Himalayas. *Contributions to Mineralogy Petrology* 136 (1999a), pp.273–284.

Schneider, D.A., Edwards, M.A., Kidd, W.S.F., Asif Khan, M., Seeber, L. and Zeitler, P.K.: Tectonics of Nanga Parbat, western Himalaya: Synkinamatic plutonism within the doubly vergent shear zones of a crustal-scale pop-up structure. *Geology* 27 (1999b), pp.999–1002.

Schneider, D.A., Edwards, M.A., Kidd, W.S.F., Zeitler, P.K. and Coath, C.D.: Early Miocene anatexis identified in the western syntaxis, Pakistan Himalaya. *Earth and Planetary Science Letters* 167 (1999c), pp.121–129.

Searle, M.P.: Extensional and compressional faults in the Everest massif, Khumbu Himalayas. *J. Geological Society London* 156 (1999a), pp.227–240.

Searle, M.P.: Emplacement of Himalayan leucogranites by magma injection along giant complexes: examples from the Cho Oyu, Gyachung Kang and Everest leuco-granites (Nepal Himalaya). *J. Asian Earth Sciences* 17 (1999b), pp.773–783.

Seastres, J.S.: Subsurface geology of the Nasuji-Sogongon sector, southern Negros geothermal field, Philippines. 4th *New Zealand Geothermal Workshop*, Auckland, 1982, pp.173–178.

Sheth, H.C. and Chandrsekharam, D.: Early alkaline magmatism in the Deccan traps: Implications for plume incubation and lithospheric rifting. *Physics of the Earth and Planetary Interiors* 104 (1977), pp.371–376.

Simpson, F. and Bahr, K.: *Practical magnetotellurics*. Cambridge University Press, New York, NY, 2005.

Sinha Roy, S., Malhotra, G. and Mohanty, M.: *Geology of Rajasthan*. Geological Society of India Publication, Bangalore, 1998.

Smith, I.E.M.: New Zealand intraplate volcanism, North Island, In: R.W. Johnson (ed): *Intraplate volcanism in eastern Australia and New Zealand*. Cambridge University Press, New York, NY, 1989, pp.157–162.

Stauffer, P.H.: Unraveling the mosaic of Palaeozoic crustal blocks in southeast Asia. *Geologische Rundschau* 72 (1983), pp.1061–1080.

Stefansson, V.: Estimate of the world geothermal potential. *Geothermal Workshop, 20th Anniversary of the UNU Geothermal Training Program in Iceland*, October 13–14, 1998, Grand Hotel Reykjavik, 1998.

Stefansson, V.: Investment cost for geothermal power plants. *Geothermics* 31 (2002), pp.263–272.

Stefansson, V.: Success in geothermal development. *Geothermics* 21 (1992), pp.823–834.

Stern, T.A.: Assymmetric back-arc spreading, heat flux, and structure of the central volcanic region of New Zealand. *Earth and Planetary Science Letters* 85 (1987), pp.265–276.

Sussman, D.: Status and geologic setting of geothermal fields in Central America, Mexico, and the Caribbean. In: R.L. Miller, G. Escalante, J.A. Reinemund and M.J. Bergin (eds): *Energy and mineral potential of the Central American-Caribbean regions*. Circum-Pacific Council for Energy and Mineral Resources, Earth Science Series, Vol. 16, Springer-Verlag, Berlin, 1995, pp.217–224.

Texas Instruments: Geothermal resources project–Stage 1. Parts 1–10, Final report, unpublished, 1971.

Thakur, V.C.: Plate tectonic interpretation of the western Himalaya. *Tectonophysics* 134 (1987), pp.91–102.

Thussu, J.L.: Geothermal energy resources of India. *Geological Survey of India Special Publication* 69, 2002.

Tobías, E.: Proyecto geotérmico Amatitlán: Estudio de prefactibilidad. Instituto Nacional de Electrificación (INDE), Guatemala, 1987.

Tole, M.P.: Low enthalpy geothermal systems in Kenya. *Geothermics* 17 (1988), pp.777–783.

Tonani, F.: Some remarks on the application of geochemical techniques in geothermal exploration. *Proceedings 2nd Symposium on Advance European Geothermal Resources*, Strasbourg, 1980, pp.428–443.

Truesdell, A.H. and Hulston, J.R.: Isotopic evidence on environments of geothermal systems. In: P. Fritz and J.C. Fontes (eds): *Handbook of environmental isotope geochemistry*. Elsevier, New York, NY, 1980, pp.179–226.

Truesdell, A.H., Nathenson, M. and Rey, R.O.: The effects of subsurface boiling and dilution on the isotopic composition of Yellowstone thermal waters. *J. Geophysical Research* 82 (1977), pp.3694–3704.

Truesdell, A.H.: Summary of section III geochemical techniques in exploration. *Proceedings Second United Nations Symposium on the development and use of geothermal resources*, Vol. 1., San Francisco, Washington, DC, 1976.

Tseesuren, B.: Geothermal resources in Mongolia and potential uses. Report 15, in Geothermal Training in Iceland 2001. The United Nations University, Geothermal Training Programme, Iceland, 2001, pp.347–374.

Ufimtsev, G.F.: Morphotectonics of the Mongolia-Siberian mountain belt. *J. Geodynamics* 11 (1990), pp.309–325.

UNFCCC: El Hoyo-Monte Galán geothermal project. United Nations Framework Convention of Climate Change, USIJI Uniform Reporting Document: Activities implemented jointly under the pilot phase. UNFCCC, Bonn, Germany, 1997, http://wwwunfccc.int/program/aij/aijact/nicusa01.html (accessed May 2004).

UNFCCC: United Nations framework convention for climate change, synthesis report. WHO and UNEP, 2006.

United Nations Development Program (UNDP) 1973: Geothermal resources development, Nicaragua. Report WR 08, 10, unpublished, 1980.

Valdimarsson, P. and Eliasson, L.: Factor influencing the economics of the Kalina power cycle and situations of superior performance. *Proceedings International Geothermal Conference*, 14–17 September 2003, Reykjavik, Iceland, 2003, pp.33–40.

Vargas, J.R.: *Geología, hidrogeoquímica y modelo conceptual del reservorio para la prefactibilidad del campo geotérmico Poco Sol, San Ramón-San Carlos, Costa Rica*. MSc Thesis, Univerity of Costa Rica (UCR), San José, Costa Rica, 2002.

Vega, E., Chavarría, L., Barrantes, M., Molina, F., Hakason, E. and Mora, O.: Geologic model of the Miravalles geothermal field. *Proceedings World Geothermal Congress*, 24–29 April 2005, Antalya, Turkey, 2005.

Vijayaraghavan, S. and Goswami, D.Y.: Organic working fluids for a combined power and cooling cycle. *J. Energy Resources Technology* 127 (2005), pp.125–30.

Vimmerstedt, L.: Opportunities for small geothermal projects: Rural power for Latin America, the Caribbean and the Philippines. National Renewable Energy Laboratory report, NREL/TP-210-25107, Livermore, CA, 1998.

Vimmerstedt, L.: Opportunities for small geothermal power projects. *Geo Heat Centre Bulletin* 20 (1999), pp.27–29.

Vimmerstedt, L.: Small geothermal projects for rural electrification. In: D. Chandrasekharam and J. Bundschuh (eds): Geothermal resources for developing countries. A.A. Bakema, Leiden, The Netherlands, 2002, pp.103–129.

Volpi, G., Manzella, A. and Fiordelisi, A.: Investigation of geothermal structures by magnetotellurics (MT): an example from the Mt. Amiata area, Italy. *Geothermics* 32 (2003), pp.131–145.

Weaver, S.D., Sewell, R.J. and Smith, I.E.M.: New Zealand intraplate volcanism, petrological overview and tectonic relationships. In: R.W. Johnson (ed): *Intraplate volcanism in eastern Australia and New Zealand*. Cambridge University Press, New York, NY, 1989.

World Bank: Rural energy and development: Improving energy supplies for two billion people. The World Bank Report, Washington, DC, 1996.

World Bank: Meeting India's future power needs: Planning for environmentally sustainable development. World Bank Report, Washington, DC, 1999.

World Resources Institute (WRI): Earth trends—The environmental information portal. On-line database, http://earthtrends.wri.org/ (accessed July 2007).

Zaide, M.C.: Interpretation of rock-dating results in the southern Negros geothermal field. Philippine National Oil Company-Energy Development Corporation, Geothermal Division Report (unpublished), Manila, 1984.

Zhao Ping, Jin Jian, Zhang Haizheng, Dor Ji and Liang Tingli: Gas geochemistry in the Yangbajing geother-mal field, Tibet. In: G.B. Arehart and J. Hulston (eds): *Proceedings of the 9th International Symposium on Water-Rock Interaction*. A.A.Balkema, Leiden, The Netherlands, 1998, pp.657–660.

Zorin, Yu.A., Novoselova, M.R., Turutanov, E,Kh. and Kozhevnikov, V.M.: Structure of the lithosphere of the Mongolia-Siberian mountainous province. *J. Geodynamics*, 11 (1990), pp.327–342.

Zúñiga Mayorga, A.: Nicaragua country update. *Proceedings World Geothermal Congress*, 24–29 April 2005, Antalya, Turkey, 2005.

Index I: Subject index

Index II: Localities, stratigraphic units, tectonic and structural elements

West Bank 37
West coast geothermal province 69
Western Sahara 37

Xizang-Yunan 63

Yangbajing 62–64, 77
  geothermal field 62–64

Yemen 22, 37
Zambia 39, 77
Zanskar region 62
Zapatera island 56
Zavkhan 72
Zebra pyroclastics 56
Zimbabwe 37
Zunil 52, 54